Sun, Wind, and Biomass

Sun, Wind, and Biomass

India's Path to a Sustainable Future

PREM SHANKAR JHA
MICHAEL B. MCELROY

OXFORD
UNIVERSITY PRESS

OXFORD
UNIVERSITY PRESS

Great Clarendon Street, Oxford, OX2 6DP,
United Kingdom

Oxford University Press is a department of the University of Oxford.
It furthers the University's objective of excellence in research, scholarship,
and education by publishing worldwide. Oxford is a registered trade mark of
Oxford University Press in the UK and in certain other countries

Published in the United States of America by Oxford University Press
198 Madison Avenue, New York, NY 10016, United States of America

British Library Cataloguing in Publication Data
Data available

Library of Congress Control Number: 2024952476

ISBN 9780198944737

DOI: 10.1093/9780198944768.001.0001

The manufacturer's authorized representative in the EU for product safety is
Oxford University Press España S.A., Parque Empresarial San Fernando de Henares,
Avenida de Castilla, 2 – 28830 Madrid (www.oup.es/en).

Foreword

This book brings together two streams of research. The first is the ongoing, worldwide search for ways to replace fossil fuels with renewable sources of energy within the very short time left if runaway global warming is to be avoided. The second is how to accomplish this objective without slowing economic growth in the poorer three-quarters of the world which has not had the chance to fully exploit the capacity of fossil fuels to improve their quality of life before being asked to give them up. The absence of a consensus among governments on how these two goals can be reconciled has been apparent in the annual meetings of the Conference of Parties to the Kyoto Protocol, especially after the Copenhagen Summit of 2009. Since then, the failure of the rich nations to meet their commitment to provide $100 billion a year of financial assistance to the poor has become, with every passing year, an increasingly acrimonious subject of discussion. We have written this book with the intention of showcasing, through a study of India, how developing countries can decarbonize their energy consumption using the three primary zero carbon sources of energy with which they are most abundantly endowed—sunshine, wind, and biomass—and to do so without sacrificing economic development, while at the same time improving the quality of the lives of their citizens.

The world was already aware of the need to reconcile these two objectives through the impact the explosive growth of China in the 1990s and 2000s had on global carbon emissions. Some 30 years ago, a programme was set up at Harvard University, the Harvard-China Project on Energy, Economy, and Environment, to study China's potential to reduce its carbon emissions without retarding its economic growth. Since then, this Project has hosted more than 250 scientists to work on this topic at Harvard, 225 from China.

The decision to extend the scope of the Project to cover India as well came as the result of a meeting in 2018 between personnel of the Project and Indian journalist and economist Prem Shankar Jha. By 2017 India had been recording an annual GDP growth of over 7% for one-and-a-half decades, and its greenhouse gas emissions, though still low in comparison to China's, were rising rapidly. Recognizing that such rapid growth would soon make India another major contributor to global warming, but that slowing its economic growth was not an option, Jha had written a book in 2017 titled *Dawn of the Solar Age: An End to Global Warming and Fear*, in which he advocated a shift from fossil fuels to selective renewable energy sources that could achieve this purpose while at the same time minimizing the impact on India's economic development.

Following an extended discussion of the economic and political impediments to decarbonization in India, the decision was made to extend the ambit of the Harvard-China Project to study how India could end its reliance on fossil fuels without retarding economic growth. An enticing feature of the challenge was the awareness that many of the strategies advanced to merge decarbonization with development in India could also be accessible to other developing countries currently relying heavily on technology and finance from the West.

That discussion was the starting point for the present book. It was conceived, and implemented subsequently, by McElroy and Jha during a series of visits by Jha to Harvard over the ensuing 5 years. Our hope is that the book will not only provide an account of ongoing research addressing the need to avert hazardous climate change, but also be an example of how the three largest countries in the world can set aside their political differences and work together to ensure the survival of the civilizations our ancestors have built since the end of the Great Ice Age, together with a steady improvement in the quality of life embedded globally in an era of peace and harmony.

Today an escalating succession of extreme weather events across the globe is ending any remaining doubt about the imminence, and deadliness, of the threat to humanity that global warming poses. Prem Jha's understanding of the history, government, and economic and social structure of his home country has made it possible for us to show how the fusion of energy decarbonization and economic development can transform the future of the second most populous country in the world, and possibly provide a beacon to guide other low-income countries out of a dependence on fossil fuels without hindering their economic development. Michael McElroy's understanding of the physics of Earth's atmosphere and his quarter century of commitment to averting the catastrophe that will result from further global warming have enabled us to provide data on the capacity of non-fossil energy to avert if not all, then the deadliest portion of it in the years that lie ahead. The imminence of this catastrophe is still not fully understood by its future victims. Therefore, in four final chapters of this book that we have placed as appendices to our study of India, we have described the deadly nature of this threat, how suddenly it has emerged, how fleeting and tenuous is mankind's control over the forces that have created it, and therefore how little time is left in which to make the changes that will allow us to maintain a livable environment for our children, grandchildren, and future generations.

Acknowledgements

Between 2018 and 2022 the Harvard-China Project has published a variety of papers that assess India's energy endowment, its diverse energy needs, the wide variations in demand and supply between its regions, how these are currently being met, and how they can be channelled to achieve a sustainable decarbonized energy system in the future. Important contributors to this research include Alex Archibald, Shi Chen, Xinyu Chen, Meng Gao, Haiyang Lin, Tianguang Lu, Xi Lu, Chris Nielsen, Peter Sherman, Shaojie Song, and Xi Yang. Their contributions are highlighted in a number of the chapters that follow. We owe a particular debt to Haiyang Lin and Peter Sherman who read early drafts of specific chapters and ultimately the entire manuscript at various stages of its production. They made important contributions not only to the structure of the book but also to its genesis. We owe a debt also to Cecilia McCormack, who patiently reconciled innumerable additions, deletions, and changes to the text made by a pair of authors located half a world and ten hours apart, while simultaneously living up to the variety of other administrative functions for which she had responsibility at Harvard.

In India, where Jha's interest in renewable energy alternatives was sparked first by OPEC's oil price hikes of the 1970s, the list of people to be acknowledged is not only long, but stretches back in time, to individuals, several of whom are no longer alive, but whose contributions were notably important. These include the late Dr M.S. Swaminathan, Deputy Chairman of the Indian Planning Commission in the early 1980s and subsequently Director General of the International Rice Research Institute in the Philippines; the late Dr R.K. Pachauri, Chairman of the Intergovernmental Panel on Climate Change (IPCC) from 2002 to 2015; and Dr Ashok Khosla, Chairman and founder of Development Alternatives and President of the International Union for the Conservation of Nature (IUCN) from 2008 to 2012. Khosla was the earliest advocate in India and, along with E.F. Schumacher, possibly in the world, for the adoption of non-energy-intensive modes of production that would bring several types of manufacturing back to village levels and generate livelihoods there. This would limit the rise of congested, increasingly unhealthy, and crime-ridden cities as a price of fossil fuel-based industrialization. Several sections of this book reflect how his writings and the advocacy-through-demonstration in his NGO *Development Alternatives* influenced Jha's thinking, and consequently the structure and contents of this book.

We would like to record also the important support Jha received from two Indian Prime Ministers. The first was Mrs Indira Gandhi, who, after hearing a 10-minute presentation on municipal solid waste gasification from Jha in October 1983, the very next month sent a study team to explore municipal solid waste gasification systems that were then being developed in the UK, the US, Japan, and Germany. The team presented its report to the Maharashtra government in December 1983, but the report died a premature death after Mrs Gandhi's assassination on 31 October 1984. The second was Prime Minister Manmohan Singh, whose Special Adviser on Climate Change, Shyam Saran, arranged for Jha to make a second visit to these countries, and additionally to Sweden in 2008, to learn first hand about the EU's Altener project, under which it had sanctioned the addition to gasoline of methanol, obtained from biomass gasification, to meet its current and future 'greening' targets.

A further seminal contribution in the 1980s was from Amitabh Pandey, a visionary Indian administrator who was the Chairman of a state government parastatal enterprise—the Punjab Ago-Industries Corporation. Pandey introduced Jha to the pioneering work of Luz, a US-based Israeli company later renamed BrightSource Energy, Inc, which set up the first commercial solar thermal power plants in California in the 1980s, and to Suresh P. Babu, the multiple-award-winning co-developer of RENUGAS®, the fluidized bed gasifier installed at the Gas Technology Institute at Chicago. RENUGAS remained the standard biomass gasifier for most research activities up to the development of the dual-chamber fluidized bed, plasma, and plasma-assisted gasification systems in the current millennium.

A special thanks goes to J.S. Samra, former Deputy Director General of the Indian Council of Agricultural Research, who generously shared his encyclopaedic knowledge of Punjab's agricultural crisis, the so far unsuccessful attempts to overcome it through the exploitation of its abundant crop residues for power generation, and their failure to do so economically, up to the time of writing of this book.

We end by acknowledging our debt to the thousands of dedicated scientists and engineers in a hundred or more countries who have worked tirelessly and selflessly in virtually every area of climate-related research to break the barrier of disbelief that had blocked research on arresting climate change for close to three-quarters of a century after the threat from global warming was first perceived. Through their combined efforts, ably orchestrated by the IPCC, they have provided the inspiration to guide how this catastrophe can be avoided.

Contents

Figures

Tables

1
Introduction

Our world is facing a battle for survival but has yet to decide where the fight must start. One hundred and ninety six countries signed the Paris Climate Agreement in 2015, committing to implement policies that would limit global warming to 2 °C, aspiring to restrain it to 1.5 °C by 2100. But well before the Glasgow Summit of 2021 it had become apparent that the world was going to overshoot both targets by a wide margin. By 2021, on the eve of Glasgow, the average global surface air temperature had risen already by 1.1 °C above the temperature in the pre-industrial age, and it had become apparent that the world was on track for a rise of as much as 2.7 °C by 2100 BCE.[1] If this trend is allowed to continue, climate change may well be out of human control by then, and we will be in the early stages of a global catastrophe likely to imperil not only our species but also much of animal, plant, insect, and marine life on Earth.

Most of the responsibility for this rests on the richer nations of the world: the US, the European Union (EU), China, Russia, India, Canada, Japan, Brazil, Iran, and Indonesia. These ten nations accounted for 68% of greenhouse gas emissions in 2021. At the opposite end of the scale, the 100 smallest countries accounted for only 3%. Among the top ten emitters, three, the US, the EU, and China, are the main culprits, for together they account for almost half, 46%, of the emissions. Although still a long way behind, India comes immediately after them, for it is now the fourth largest greenhouse gas emitter in the world.[2]

By the start of the Glasgow Summit, nine of the ten largest emitters had set target dates for when they would attain net zero emissions. India was the lone exception. So, on 1 November 2021, when Prime Minister Narendra Modi announced that India would attain 'net zero' carbon emissions status by 2070, he electrified the conference. To demonstrate

Sun, Wind, and Biomass. Prem Shankar Jha and Michael B. McElroy, Oxford University Press.
 DOI: 10.1093/9780198944768.003.0001

his seriousness, Modi also committed India to meeting four intermediate targets by 2030.

These were to generate 50% of India's electricity from renewable energy; increase its renewable energy generating capacity to 500 gigawatts (GW); reduce India's overall carbon emissions by 1 billion tonnes; and reduce the carbon intensity of its energy use per unit of GDP by 45%. But he did not give even a hint, let alone spell out *how* India proposed to meet the targets he had announced. In this he was not alone, because none of the other 196 countries' governments present at the Summit had done so either. The industrialized countries had encouraged scientists to find and develop renewable energy technologies, but then left it to their market-driven economic systems to decide which technologies would succeed and which would perish by the wayside. The scientific community did not let the world down.

By the Copenhagen Climate Summit, 215,000 patents had been filed that focused on renewable energy technologies.[3] A study by the World Intellectual Property Organization (WIPO) found that 77,813 patents relevant to renewable energy had been filed in the first decade of this century. Of these, the majority were filed in Japan (55%), followed by the US (20%) and Europe (9%).[4] But leaving it to the market to decide which technologies would succeed and which would fail took both the power to choose between the various renewable energy technologies on offer and the power to determine the speed of the shift out of fossil fuels from the hands of the state transferring the responsibility largely to the private corporate sector.

Since the corporate sector has conflicting interests and lobbies, this has slowed the pace of transition at a time when speed is the essence for this challenge the world faces. The principal cause is the struggle between established and new technologies, which economist Joseph Schumpeter labelled the process of 'Creative Destruction'. For more than a century, as the pace of technological change has quickened, this struggle has increasingly defined the modern market economy.[5] Today it is doing so in the energy sector, where fossil fuels that have been the dominant source of commercial energy for 200 years are staging a rear-guard battle to protect their market domains against the new energy technologies being developed under the spur of climate change. As a result, renewable energy technologies are having to compete against fossil fuels that, according to the most recent International Monetary Fund (IMF) study, continue to benefit from $490 billion a year in explicit subsidies. Most of the recipients are headquartered in the 36 industrialized countries that are the members of the Organisation for Economic Co-operation and Development (OECD) and their eight large partner economies—Argentina, Brazil, China, Colombia, India, Indonesia, Russia, and South Africa.[6] Not by coincidence, these are precisely the countries responsible for more than three-quarters of the world's greenhouse gas emissions today.

Since the sole criterion that the market uses in deciding investment is private profit, i.e. returns to the investor as distinct from returns to society, *this virtually rules out any consideration of the consequences for society as a whole of the choices it makes.* During the era of national capitalism and the nation state, the balance between the investor and society was maintained by the development of trade unionism and the welfare state. But even during that era, it took a long time to recognize the importance of government action to protect the weak. This happened first in Germany, in the 1870s under the chancellorship of Bismarck, a full half-century after the start of German industrialization. In Britain, it took until 1909, almost a century after the Luddite uprising of 1811, the first protest against industrialization, for parliament to accept the need to create a welfare state.

Today, the imperative need to stop using fossil fuels is recognized by all governments, but action to do so has encountered the same inertia that delayed the creation of the welfare state. The need for the international community to tilt the balance of state incentives towards renewable energies was recognized explicitly at the Copenhagen Summit, but a commitment made then to create a $100 billion a year fund to help the developing countries shift out of fossil fuels had still to be implemented by the end of 2021, a full 12 years later.[7]

In the OECD and partner countries, beyond providing a few limited incentives for reducing carbon emissions, such as those provided under the Clean Development Mechanism, no consensus has emerged among the major energy-consuming countries on the need for concerted global action and for a strategy to discourage fossil fuels and encourage their replacement by renewables.

If the world had a century in which to attain 'net zero' it could have left the shift from fossil to renewable energy to the market. But we have at most a few decades, and the change has to be made against the steeply rising slope of not merely continuing government subsidies on fossil fuels, but continuing investment in the development of new technologies for extracting fossil fuels, of which fracking is the latest, most lethal, example. The need for decisive government action to break down this structure of subsidies and provide strong incentives for the development and use of renewable energy has therefore never been greater. This requires the state either to fight the entrenched power of the major fossil energy producers who dominate the energy market or, which would be preferable, to give them a sufficient incentive to shift out of fossil fuels into renewables much sooner than they would otherwise. Continuing to leave the timing of the shift to the market is a recipe for disaster.

This is particularly true of transport fuels, as discussed below, sudden, giant fluctuations in oil prices have made investment in biofuels virtually impossible without remunerative price guarantees from governments.[8]

Creating a stable and remunerative market environment for new technologies requires therefore meeting not one but three sets of interrelated challenges. The first is the development of as wide a range of technology options as possible. As stated above, this has been more or less achieved. The second is selecting the renewable energy technologies that go furthest towards replacing fossil fuels without impeding a country's economic growth, employment generation, and enhancement of goals for its well-being. The third is that since renewable energy generation, unlike fossil fuels, requires a lot of land, and most of this land is already populated and committed to other uses, each country's government will need to frame pricing, taxation, and subsidization policies and additional incentives to promote willing cooperation by the owners and users of this land.

Humanity's consumption of energy is so vast today that the most important source capable of meeting society's needs is the sun. The Earth receives its potential sustainable energy directly as light and heat from the sun, and indirectly through derivative sources of wind and biomass. Capturing and concentrating the heat of the sun through solar concentrators and using this to replace coal and gas in power generation was the first alternative source to be developed. This occurred in California in the early 1980s and received a great fillip when scientists, a decade later, developed ways of capturing and storing the heat of the sun in the form of two molten salts, potassium and sodium nitrate. This made it possible to harness round-the-clock power from the sun, beginning in the late 1990s.

But in the warm countries of the South, solar thermal lost out to solar photovoltaic (SPV) power, when the cost of solar panels dropped abruptly in the second decade of this century following major breakthroughs in technology that increased their efficiency and the entry of China into the global market for solar panels.

Wind power generation has developed equally rapidly, and its cost of generation has fallen only a little less rapidly over the past decade. These two sources have therefore developed commercially to the point where their 'base' or 'overnight' cost of generation has fallen below even the heavily subsidized cost of power generated from coal and natural gas. But these technologies have still to surmount the challenge of reliably satisfying demand as required at any time of the day or night—features that coal and gas are meeting today.

Technologies for producing transport fuels from biomass were developed in the US, Japan, Germany, and the UK as far back as the 1990s, but failed to be commercialized due to the extreme volatility of oil prices that followed the onset of the economic recession in 2009. This brought oil prices crashing down from $140 a barrel in mid-2008 to below $40 in mid-2009. They climbed back to $87 in 2010, but were brought down again by the US exploitation of fracking, to $51 by January 2020. Interest in these technologies was revived in 2022 after the sharp rise of oil and natural gas prices triggered by the Russo-Ukrainian war.

In the USA the first commercial plant designed to produce biofuels from municipal solid waste went on stream outside Reno, Nevada, and began to produce synthetic aviation fuel (SAF) from December 2022. Unfortunately this plant ran into teething troubles a year after it began, but before it could raise the funds it needed to resolve them and re-start the plant, a group of US senators demanded that the US government withdraw incentives in place for non-renewable feedstock sources under the Inflation Reduction Act. Non-renewable feedstock was defined to include landfill waste.[9]

There are no one-size-fits-all solutions

Today, the world has met only the first of the three challenges mentioned above. It has made virtually no headway in meeting the second—selecting the renewable energy technologies best suited to each country's needs. And, it has made absolutely no progress to meet the third. Contrary to the impression that has been given by the media, there are no 'one size fits all' solutions for the whole world. Each country, or region, has to find the right mix of technologies on its own, reflecting intrinsic differences in their endowments for alternatives to replace fossil fuels. Thus, while northern European countries and the northern part of the US have an abundance of wind energy but relatively few hours of direct sunlight, countries in the equatorial and subtropical belts have the opposite endowment. This difference in endowment is an important reason why the harnessing of SPV power has taken the form mostly of rooftop installations in the 'North', but grid-connected commercial power in the 'South'. It is also the main reason why the cold countries have not shown much interest in the development of biomass energy—for there is little or no growth of biomass there during the winter. The warm countries of the Global South do not need to follow in their footsteps because they have year-round growing periods, have mainly agricultural economies, and have therefore a relative abundance of biomass.

Land and the human factor

The third and perhaps most important requirement for speeding up the shift out of fossil fuels springs from the need the new technologies have for large amounts of land.

Today, outside of the great hot and cold deserts of the Earth and the ice-covered regions of Greenland and Antarctica, there is hardly a square mile of land on the planet that is not already occupied and providing sustenance to human beings. To get people to shift from traditional uses of their land to completely new uses will require offering them fiscal and monetary incentives that will make those affected by the change of land use its gainers, not losers—partners in the development of renewable energy, not its victims. This will happen only if they see an immediate gain from the shift. By the same token, policy measures that impoverish them, such as requisitioning their land for solar thermal or photovoltaic utility-scale plants, or placing restrictions on its future use to ensure an uninterrupted flow of wind to wind turbines, will be resisted.

In the chapters that follow we discuss how India can meet the triple challenge we have outlined here. But the conclusions we arrive at are relevant also to other developing countries and to the industrialized countries as well. This is because, despite incentives such as the Clean Development Mechanism offered by the OECD and other incentives promoted by individual developing country governments, by 2019 reliance on the market had enabled only 9% of the global demand for electricity to be met by variable renewable energy sources, i.e. solar and wind energy.[10] Large hydro and nuclear power met another 30%. Close to 60% was still generated from fossil fuels. Even these data do not fully convey the gravity of the threat the world faces. For, the annual emission of CO_2 and other greenhouse gases is still rising, and just about all of the additional contribution to greenhouse gas emissions is taking place now in the low- and middle-income four-fifths of the world, where energy consumption is increasing by more than 4% a year as compared to 0.3% per year in the OECD countries. There is thus a long way to go, and most of the road lies in the non-OECD countries.[11]

The challenge that India and for that matter all other low-income countries face is one of making this shift without sacrificing the growth of GDP and employment. This has been readily conceded in a report of the US Center for Strategic and International Studies (CSIS), which commented just before the Glasgow Summit that 'India is in the eye of the energy transition storm. *No other country must balance the challenges of implementing so many power sector reforms while expanding energy access to hundreds of millions of people*'.[12] Energy access, needless to say, is the first requirement for economic growth.

Once we concede that India, like other low-income countries, cannot afford to forego growth in favour of decarbonizing its energy supply, the challenge of reaching 'net zero' emissions, even by 2070, becomes even more daunting: the most recent estimate of the International Energy Agency is that India still meets 89.1% of its commercial energy needs from coal, oil, and natural gas and another 5.8% from nuclear and large hydroelectric plants. In 2019–20, wind and SPV energy met only 4.5% of its commercial energy needs. This will need to rise to at least 80% by 2070, even if nuclear and hydroelectric power retain their present share of energy supply, and considerably higher if they do not. What is more, if economic growth is not to be sacrificed, renewable, zero carbon sources will have to fuel as much as 80% of a GDP that could be eight to ten times larger than it is today. Some of that challenge will undoubtedly be met by increases in the energy efficiency of production. But most of it will require additional energy generation. This makes the challenge the developing countries face radically different from that faced by the industrialized countries.

An asymmetry of objectives

The OECD countries, and especially its European and North American members, are industrialized and enjoy high per capita incomes. Despite the deindustrialization that set in with economic globalization in the 1970s, they still enjoy something close to full employment. They also have social welfare systems to cushion the extremes of destitution for those who are no longer able to work or find jobs. Furthermore, since the 1980s, economic globalization has shifted the locus of industrial production from Europe and North America to Asia, reducing the tax base in the countries on these continents, subjecting their social welfare systems to significant pressure. In them, unemployment has ceased to be a cyclical product of economic recession and has become chronic. With this, trade unions have lost much of their bargaining power. Job security, pensions, and health and retirement benefits have dwindled even for mid-level workers and managers, and income inequalities have increased dramatically. At the same time, the computerization of book-keeping, inventory management, and a host of other such functions has made a large section of the white-collar workers redundant. These changes have increased political stress within governments in the industrialized countries and have begun to undermine the legitimacy of democracy in the developing countries. The industrialized countries cannot therefore be blamed for wanting to make the shift of their energy base with the absolute minimum of disturbance to their existing socioeconomic structures. The easiest way to do this is to leave both the choice of renewable energies technologies and the timing of the shift into them to the market.

The low-income countries have the opposite problem. Since the end of the Second World War, when they began the race to industrialize, they have depended on the industrialized countries to provide the technologies they need. Today, they are continuing to do so despite the increasing asymmetry between their domestic policy goals and those of the industrialized countries. In the industrialized countries there is no compelling need to increase GDP, because the problem their economic and social systems face is not one of insufficient income, but an increasingly unequal distribution of income. Some GDP growth is necessary to offset the creeping displacement of labour by machines caused by technological progress, to maintain near-full employment, and to reduce the burden of social welfare commitments on the treasury. But beyond this minimum, the principal goal of state policy in developed economies is to maintain and safeguard the quality of present life.

By contrast, in the developing countries, economic growth is, and has to remain, the primary objective of the state. All of them have rapidly growing, and therefore young, populations in search of livelihoods. Providing them with jobs is thus the first, and overriding, obligation of their governments. So far, industrialization is the only route they know. So all have been treading down this path.

Today, 70 years or so after the hectic rush to industrialize began, the road has all but closed, and nations are having to contend with multiple unintended consequences. These include rapid, uncontrolled, urbanization; rapid deforestation; rapid degeneration of their soils; accelerating desertification; galloping air and water pollution; and a multiplication of diseases for whose treatment the victims do not have the means to pay.

Being late starters in industrialization, they also face a social and humanitarian problem that the first industrializing countries did not have to confront: this is the displacement of the *existing* artisanal and technologically primitive industries of the pre-market economy, and consequently a massive deindustrialization occurring side by side with modern industrialization. This displacement was gradual in 18th- and 19th-century Britain and Western Europe because the pace of technological development was slow. The US did not even have to face this displacement because it had an endless supply of land west and south of the rapidly industrializing north-east, to which those displaced by technological change could migrate in search of a new life.

Today's industrializing countries are facing this dislocation in a much more extreme form. For, while the displacement of artisanal and backyard enterprises is far more sudden and complete, the new technologies are so advanced and so capital intensive that they require very few workers. The later that a country has to face the challenges posed by modern industrialization, the larger the jump it has to make from the traditional to the modern technologies they have to master in order to compete effectively on the world market. So the greater then has been the displacement of labour and artisanal enterprise with which they have had to cope.

Low-income countries face therefore a triple challenge: they have to find the capital for building a modern infrastructure and setting up modern industry; they have to educate a young population to the level required by these technologies; and they have to provide alternative sources of livelihood for those displaced by the withering of traditional enterprise. These are daunting challenges, so it is not surprising that, three-quarters of a century after the Second World War ended, less than a dozen of the more than a hundred late starters in industrialization have succeeded in breaking through the twin barriers of poverty and the entrenched market power of the countries that preceded them. The rest are suffering to differing degrees from what Harvard economist Dani Rodrik identified, in the 1990s, as 'stalled industrialization' (in private communication). In them, growth in the share of industry and minerals has stalled at 18–25% of GDP. This has forced their governments to look for, or live with, other routes to prosperity. Mineral-rich countries, from the Middle East to Bolivia, have found their salvation in the mining of everything from oil to lithium. Many small and island nations have successfully capitalized on tourism and financial intermediaries. Still others, ranging from Mexico and Colombia to Thailand, Myanmar, Afghanistan, and Kosovo, have found the answer in narcotics. But despite all these and other stratagems, only a few of the late starters have created a permanent, stable base for prosperity. Not surprisingly, as economic globalization has destroyed the last protectionist barriers that developing countries could have used to gain a bit of time, more and more countries have slipped into chaos, or are putting their despairing trust in dictators of either the extreme Left or Right.

It is against this difficult background that India needs to exercise its renewable energy options. It cannot afford to let the market be the sole determinant, because the market only uses the yardstick of private costs and returns. It needs to select technologies that will not only check the ballooning consumption of fossil fuels, but also enable it to do so *without preventing the future growth of income and employment.* To do this, India needs to make its own estimates of the costs and benefits of specific renewable energy technologies and not rely only on those presented to it by the vendors in the global market. These estimates need to take into account the impact that various renewable energy technologies will have on the bottom line not only of the investors but also of society.

In this book we attempt to show that an optimal combination of technologies backed by a pricing and incentive system that maximizes social benefits and economic growth can not only enable India to achieve net zero emissions status by 2070 and prospectively a good deal earlier, but also increase the growth of employment and GDP, reverse deforestation, and at the same time offset air and water pollution and counter the effects of soil exhaustion.

Chapter 2 addresses the rise and evolution of civilization in India, from that of the Indus Valley 6,000 or more years ago up to the present, taking up the issue of why India missed the opportunities for industrialization in the 19th century in spite of the fact that it was richly endowed in fossil fuels, specifically coal. Chapters 3–5 outline the current state of the Indian economy, the stalled condition of its industry, and the deepening crises in its agriculture and employment. Chapter 6 describes the dawn of awareness in India of the threat that global warming poses to its future. Chapter 7 describes the fatalistic belief, which lingered until 2017, that the country's dependence on coal and imported crude oil can only be reduced, not eliminated, if it is to realize its ambitions for economic growth, a misconception that was finally laid to rest by the recent rapid increase in extreme weather events leading to Prime Minister Modi's robust acceptance of the challenge posed not only for the world but also for India by climate change. Chapter 8 outlines India's successive assessments of its potential for renewable energy generation and its so far hit-and-miss attempts to limit its carbon emissions. Chapters 9 and 10 describe projections of India's potential for power and process heat generation from wind and solar energy and for production of green hydrogen (hydrogen obtained from renewable sources of energy rather than from coal, oil, or natural gas). Chapter 11 brings together the findings of the previous two chapters, arguing that a combination of solar and wind energy complemented by a supply of green hydrogen could bring India close to a zero carbon future as early as 2050, supplying the bulk of the energy the country needs to satisfy its demands for electricity, transport fuels, and process heat.

Gasification of biomass could make an important additional contribution to this zero carbon future, as discussed in Chapter 12. Chapter 13 reminds readers that with the turn from fossil fuels to solar and wind energy, land becomes a critically important factor of production once more, as it had been for 50,000 years before the industrial revolution. The chapter estimates the land area that is potentially available for solar and wind power generation in India, concluding that with the right mix of policies and incentives this would be sufficient to harness most of the solar and wind power potential identified in Chapter 9.

Chapters 14 and 15 describe how solar, wind, and biomass energy can resolve many of the crises endemic to both dryland and irrigated farming. The concluding chapter, Chapter 16, summarizes findings as to the increases in income, employment, environmental benefits, and economic growth that could be realized as a result of the shift from fossil to solar, wind, and biomass sources of energy.

The Appendix includes four chapters that provide a general introduction to the history and nature of the climate system. Chapters A1 and A2 outline the evolution of Planet Earth up to today from its formation 4.5 billion years ago; the evolution of human beings from *Homo erectus* to *Homo sapiens*; the birth of civilization, its subsequent development, and its gradual acquisition of the power to change the Earth's, and its own, future. Chapters A3 and A4 describe the change in the world's atmosphere that we have wrought and how, precisely, this has changed our climate to the point where our current civilization, if not our species, is facing extinction.

Notes and References

1. UN Environment Programme (UNEP) Emissions Gap Report, 26 October 2021.
2. Ibid.
3. IRENA. https://www.irena.org/errors/404?item=%2finspire%2fintellectual-property-rights%2fglobal-patenting-trends&user=extranet%5cAnonymous&site=irena-azure-cd (viewed 4/1/22).
4. https://www.un.org/en/climatechange/net-zero-coalition.

5. Schumpeter, Joseph A. (1911). *Capitalism, Socialism and Democracy*. First published 1942 by Harper & Brothers.
6. The OECD Inventory of Support Measures for Fossil Fuels, 2019. https://www.oecd.org/fossil-fuels/data/OECD-Fossil-Fuels-Support-database-brochure-2019.pdf (viewed 4/1/22). These are the combined subsidies of the OECD and eight partner economies—Argentina, Brazil, China, Colombia, India, Indonesia, Russia, and South Africa. The OECD component reached a peak of $233 billion in 2013, but fell to $140 billion in 2017. In the partner economies there was a similar downward trend after 2013 caused by the deregulation and rationalization of prices. But this too was reversed in 2017. The subsidies of the partner economies were $200 billion in 2017 (all estimates are at 2017 prices).
7. https://www.nature.com/articles/d41586-021-02846-3 https://www.oecd.org/climate-change/finance-usd-100-billion-goal/.
8. Kolakowski, Mark (updated 13 November 2024). History of Oil Prices. https://www.investopedia.com/history-of-oil-prices-4842834 (viewed 6/1/22). The price of oil is adjusted for inflation using the headline Consumer Prices Index (CPI) and is shown by default on a logarithmic scale.
9. Waste-to-SAF firm Fulcrum Bioenergy files for Chapter 11 bankruptcy. https://www.safinvestor.com/news/145909/fulcrum-bioenergy/#:~:text=In%20some%20cases%2C%20the%20sludge,nightmare%20didn't%20end%20here.
10. International Energy Agency (IEA): Global Energy Review 2020. https://www.iea.org/reports/global-energy-review-2020/renewables.
11. IEA: Electricity Information. https://www.iea.org/reports/electricity-information-overview.
12. https://www.csis.org/analysis/meeting-indias-net-zero-moment?gclid=EAIaIQobChMItueF5qrz- QIV7ZBoCR1X_Q6PEAAYAyAAEgKwm_D_BwE.

2
India's Goals and Constraints in the Battle against Global Warming

India is a vast country. In land area it is the seventh largest in the world, covering 3.4 million km^2, far below Russia's 16 million, trailing also the US, China, and Canada each with close to 10 million km^2. But size need not be measured by land area alone. A more relevant index is arable, and therefore habitable, land. There India ranks first, with 156.5 million hectares (mha), followed by the US with 153.3 mha, Russia with 123.1 mha, China with 118.9 mha, and Brazil with 80.9 mha. Canada comes much further down the list with 43.8 mha.

India forms the bulk of the land mass of what is now called South Asia. This land mass was created by the collision of the Indian tectonic plate with the Eurasian plate 51 million years ago. In the collision, the Indian plate dived beneath the Eurasian but, in the process, pushed the edge of the Eurasian plate up to form the Himalayas, by far the highest mountain range in the world. Over millions of years, the snow and ice formed on these ranges created a vast riverine drainage system that, in turn, created the Indo-Gangetic Plain, which stretches 2,000 miles in a great arc from east to west across Bangladesh, India, and Pakistan, ranging from 150 to 400 miles across from north to south.

Watered by snow melt and 3 months of monsoon rains in summer, 6,000 years ago this was the single largest stretch of fertile alluvial soil in the world. If one adds the 7 mha of arable land of Bangladesh and the 31 mha of Pakistan, today the Indo Gangetic Plain contains one-seventh of all the arable land in the world and a far higher proportion of its best part. Inevitably, over the more than ten millennia since the retreat of the Younger Dryas 'little' ice age, it has attracted wave after wave of humans from the colder, arid regions of the north and north-west beyond the Himalayas, moving in search of sun, water, pasture, and tillage. This

Sun, Wind, and Biomass. Prem Shankar Jha and Michael B. McElroy, Oxford University Press.
 DOI: 10.1093/9780198944768.003.0002

accounts for the origins of India's mind-boggling contemporary ethnic diversity.

The fusion of the genetic strains of these waves of migrants took place through conquest, intermarriage, and acceptance of cultural diversity. A seminal, decade-long DNA analysis of 132 Indian samples, encompassing 560,123 single-nucleotide polymorphisms (SNPs), confirmed in 2015 the existence of two distinct ancient gene pools, which it calls the Ancient North Indian and Ancient South Indian gene pools. It further distinguished within the two strains: the first and more numerous, an Indo-Iranian strain originating on the Iran plateau; and the second, a strain originating further west in the Caucasus region that the analysts identified as Indo-European. It was the latter that brought Sanskrit to India, whose similarities with several European languages, notably Greek and Latin, had been noticed as far back as in the 18th century.

Where the research broke new ground was in its study of uniparental markers. This revealed the existence of a common, deep-rooted *female* gene pool with 60% of females sharing the same group of DNA markers *across all the samples*, i.e. across the entire country. The male gene pool, in contrast, had multiple origins, both indigenous and from outside the subcontinent. These are strongly correlated to the language family to which the males belong. The conclusion the study leads to is that there may have been an indigenous human species in North Western India, which may *not* have arrived through migration from the Fertile Crescent, and that this strain intermingled through marriage and cohabitation with several later migrations from the Central Eurasian and Iranian Plateau regions. Today this indigenous DNA pool is concentrated mainly in Dravidian language speakers, i.e. in Southern India.

While the study does not explicitly link its analysis to migration patterns within India, it discredits the view first expounded in the 1940s by the British archaeologist Sir Mortimer Wheeler that the Indus Valley civilization, also frequently referred to as the Harappan or Mohenjo-Daro civilization (after the discovery of its first two grand cities, 650 km apart in the Indus Valley), was destroyed by an Aryan invasion circa 2600–1900 BCE (Figure 2.1). Instead, it tilts the scales strongly in favour of the far more plausible hypothesis that the Indus Valley civilization, whose origins date back beyond 3500 BCE, eventually fell victim to climate change.[1]

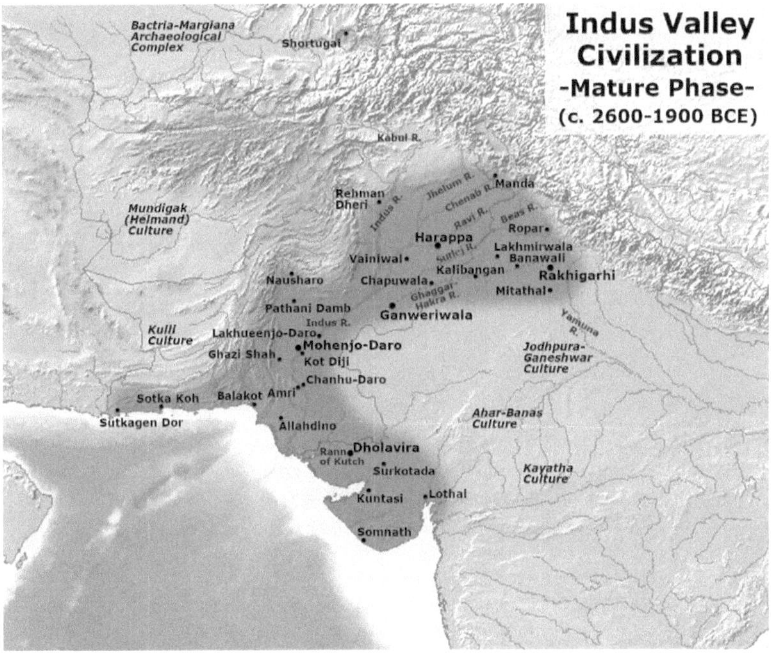

Figure 2.1 Area occupied by the Indus Valley civilization.

Source: McIntosh, Jane (2008). *The Ancient Indus Valley: New Perspectives* (ISBN 978-1-57607-907-2). https://en.wikipedia.org/wiki/Indus_Valley_Civilisation (viewed 12/1/2022) (https://creativecommons.org/licenses/by-sa/3.0/).

Archaeological evidence collected over several decades in the 20th century shows the existence of a highly developed civilization, at least comparable, if not superior, in wealth and technological sophistication to the contemporaneous Sumerian and Egyptian civilizations. According to Sumerian texts, in that world of the first great civilizations, the Indus Valley civilization was known for its technological sophistication, its extraordinary levels of standardization, and the skill of its artisans. Cities such as those found at Mohenjo-Daro, Harappa, Lothal, and the most recently discovered Dholavira included multi-storeyed rectilinear buildings arranged on a grid plan, as illustrated in Figure 2.2. They also had indoor toilets and covered drains for sanitation and large water reservoirs, which suggested a need for conservation, as the world warmed and dried out after the end of the Younger Dryas around 6000 BCE.

Figure 2.2 View of some of the ruins of Mohenjo-Daro, showing three-storey buildings.
Source: https://upload.wikimedia.org/wikipedia/commons/2/2b/Mohenjo-daro.jpg.

Three features distinguished the Indus Valley civilization from the Sumerian and Egyptian. First was a high degree of centralization of administration. Second, despite the civilization being spread over close to a thousand kilometres on both sides of the then verdant Indus Valley, and cities such as Harappa and Mohenjo-Daro separated by as much as 650 km, all were built with five identical sizes of baked clay bricks. Harappa had six rows of granaries estimated to be capable of feeding 80,000 persons or more. What is even more telling is that the granaries had brick-lined air ducts built into the floors to allow warm air to be pumped into the stored wheat to keep it dry and edible. Third, the society included artisans with very high technical skills. Sumerian texts laud the weights and measures developed in the Indus Valley as the most accurate and uniform, indicating that they were in wide use there, and possibly also in Egypt. These texts refer also to the Indus Valley people as the jewellers of the ancient world because of their unparalleled capacity to work precious stones finely with soft bronze instruments, at a time when the Iron Age had yet to dawn and iron and steel implements were unknown. A highly celebrated woman's girdle belt composed of parallel strings of precious and semiprecious stones, displayed by the Archaeological Survey of Pakistan at an Asia Society exhibition in New York in February 1998, demonstrated this extraordinary skill. The belt contained parallel strings of carnelian beads 5–6 cm long, but only 0.3 cm in diameter. To string these together into a girdle, the jeweller had to know how to make a hole lengthwise, through the centre of a carnelian bead 5 cm long but only 3 mm in diameter. This skill apparently did not exist in either Egypt or Sumeria.[2]

The excavations at Mohenjo-Daro also conclusively discredited Mortimer's Aryan invasion theory because they showed that many of the bodies found in the eighth, and highest, layer of the city were of people who had drowned. This and the existence of large water reservoirs in all the major cities suggest that the Indus Valley had begun to suffer from the periodic droughts and floods that go with a warming of the planet.

Two theories have been advanced to explain the end of the Indus Valley civilization on the basis of climate change. The first is that one, possibly two, of the eastern tributaries of the Indus ran dry as the western sides of the Indian subcontinent dried out, as they had done in other parts of the world after the end of the previous Ice Age. In the Indus Valley this would have happened after the end of the Younger Dryas mini-ice age. The drying out of two eastern tributaries of the River Indus almost certainly forced some of the inhabitants to move east into the Gangetic Plain. The second possibility is that they sailed down the Indus southwards to the Arabian Sea and then proceeded south along the coast, settling vacant arable land wherever they found it. In more recent times, the second hypothesis has gained credence over the first because most of the minerals and precious metals used in the making of the jewellery, which was one of its principal exports, were being obtained from mines in Karnataka and further south.

This has been confirmed by the DNA studies described above, which show a concentration of the 'indigenous' gene pool in the Dravida language speaking areas of the country. This suggests that the majority of the exodus from the Indus Valley was to the south, with settlements developing around river mouths, until the migrants turned round the southern tip of the subcontinent and found the broad, well-watered plains of northern Sri Lanka, Tamil Nadu, and possibly, at a later date, the Mahanadi River Delta in Odisha.

Finally, there was a fourth feature of the Indus Valley civilization that made it very different from those of Egypt and Sumeria. This was the absence of war. No weapons were found in any of the lower seven layers of the excavated buildings of Mohenjo-Daro. A few spears, swords, and suchlike were found only in the topmost (eighth) layer, but archaeologists were unable to decide whether these were meant for defence or for hunting.[3]

The origins of ethno-federalism, and how a nation state was built without war

Modern India would never have existed had it not been for its people's easy acceptance of ethnic diversity. This arises from the fact that while there have been many empires both in northern and southern India, the entire country has come under one paramount ruler only three times over the past 2,500 years for a total of 450 years: 140 years during the Mauryan dynasty (322–185 BCE); 160 years under the greater Mughals (roughly 1550–1708); and 150 years from the end of the 18th century until 1947 under the British. But there was a much longer period during which major parts of the country, whether in the north or the south, existed in peace and prospered under a single paramount ruler.

Thus, the Nandas, who were the predecessors of the Mauryan dynasty, and the Shungas, who succeeded it, ruled most of northern and central India from Pataliputra (present-day Patna) in Bihar for almost 400 years. Similarly, the Mughals had been preceded by several Afghan dynasties that ruled northern India from Delhi for 300 years before the Mughals arrived in 1526.

A similar empire, the Vijayanagara Empire, depicted in Figure 2.3, fabled for its wealth and military prowess, sprang up in the Deccan in the 14th century, extended its paramountcy across southern India and Odisha on the east coast, and kept the Muslim Empire and its principalities at bay beyond a fluid border region bounded by the Tungabhadra and Krishna Rivers until well into the decline of Mughal power in the late 17th century.

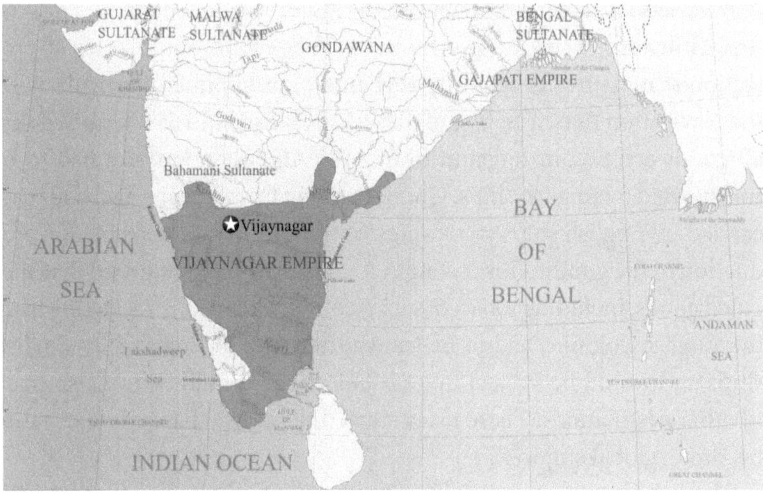

Figure 2.3 Vijayanagara Empire map, Vijayanagara dynasty.
Source: https://en.wikipedia.org/wiki/Vijayanagara_Empire#/media/File:Indian_Empire_of_Vijaynagar_map.svg.

Arrival of the Europeans and the beginning of colonisation—how one of the earliest and richest human civilisations became one of the poorest

The Portuguese explorer Vasco da Gama was the first European to successfully negotiate a direct sea route from Europe to India in 1492. Da Gama landed in Calicut (now Kozhikode), on the south-west coast of India. To his surprise, he found there not a primitive society that could be induced to part with its treasures, particularly its spices, for a few Portuguese trinkets, but an advanced, savvy community of merchants and artisans from the African coast, South-east Asia, and China. The European explorers, notably the Portuguese, were late arrivals in India. But they differed from their predecessors because they were interested from the outset not only in trade but in establishing a permanent presence in India for purposes of trade, and later conquest. Conquest began with the arrival of an expedition led by Pedro Cabal, leading a flotilla of thirteen ships and twelve hundred men. Cabral established a durable, and continuing role for Portugal in the lucrative eastern spice trade. Catholic Spain and Portugal lost out eventually, however, to France, Holland and notably England.

A variety of factors contributed to this British ascendancy. The first was its investment in a powerful navy. The second was a series of protectionist measures enacted by parliament, exemplified particularly by the Navigation Acts. The first of these Acts, passed in 1651, required that all goods arriving in England from Asia, Africa, or America had to be transported by English ships. Goods arriving from Europe also had to be carried by English ships or by ships of the country responsible for production of the goods. This second Act, passed in 1660, required that specified goods, including tobacco, sugar, cotton, and wool, originating from the English colonies had to be transported first to England by English ships and would be subject to a tax before they could be trans-shipped to other destinations. These taxes created the capital base of what was to become a global empire.

The initial contact of Britain with India was limited to trade, which was entrusted to the East India Company. The Company took advantage of the weakening of the Muhal empire after the death of Aurungzeb, the last of the Great Mughal rulers of India, overthrew an inept and cash strapped Mughal Governor of Bengal at the Battle of Plassey in 1757 and defeated a more powerful challenge from the Mughal ruler of Delhi in 1770 to consolidate the first permanent British enclave in eastern India. From there, with the help of short-lived alliances with a succession of suddenly independent local rulers that were the detritus of the Mughal Empire, the East India Company, now a virtual private army of Britain, expanded its control to virtually the whole of north and south India, in the next 80 years. By the end of the 19th century Britain ruled the entire subcontinent, just over half of it directly and the other half through subservient princes guided by a Crown's representative.

The British, who came first as traders and developed colonial ambitions only later, adapted readily to the principles of paramountcy. At the time of their departure in 1947, they administered 52% of the country directly. The remaining 48% was administered by 562 princes, advised by British political officers who oversaw the implementation of British policy in the states but also represented their interests to the British official hierarchy in the capital.

When the freedom movement, which started with the creation of the Indian National Congress in 1885, began to gather steam in the first years of the 20th century, it unconsciously replicated this decentralized pattern.

The Congress party that gained India its independence in 1947 was a coalition of freedom fighters heading local ethno-national movements for independence from the British, or from their own princely rulers, bound together by the charismatic personality of Mahatma Gandhi and his commitment to win independence without having to resort to violence.

This led not only to the most graceful exit of a colonial power from any former colony, but also to the preservation of the ethno-linguistic diversity of the subcontinent. The Indian Constitution gave formal shape to this in its first article, describing India as 'A Union of States', implicitly conceding that the states had pre-existed the new Union of India. This was how the world's largest modern nation state was created not only as a democracy, but also, in the sharpest possible contrast to the creation of nation states in Europe and most other parts of the world, without resort to war.

In April 1994, during a state visit to the US by the Indian Prime Minister P.V. Narasimha Rao, after their one-to-one meeting at the White House, President Clinton described India's achievement to the assembled journalists as 'amazing'. As a Rhodes scholar and student of history, he was perfectly aware that India had done what no other country had managed to do to this point.

How India became a late starter in industrialization

Since India's independence, more than 130 former colonies have also sought independence in different parts of the world, but very few have succeeded without having to fight for it. Those wars of independence extracted a huge price from the people in many countries, one that several of them are still paying. But India was spared the burden of having to heal the psychological wounds and repair the damage independence movements have had to suffer elsewhere. But what it was not spared, and what it has had to contend with virtually from day one, is the denial of the opportunity to industrialize, which its status as a colony imposed upon it for an entire century after the failure of the uprising for independence in 1857 (called the Great Mutiny, by the British).

These lost years made India a late starter in industrialization, when it could easily have been among the earliest. Over the two centuries when a quarter of the world grew richer, unravelled the secrets of the universe, and developed an enviable lifestyle that no prophet of antiquity could have predicted, India, once one of the two richest countries in the world, found itself left far behind. Its status as a colony was partly responsible, but many of the wounds India has suffered have been self-inflicted. These wounds are the subject matter of this book.

Up to the end of the 18th century, while mankind relied for its energy needs on animal, water, and wind power the distribution of wealth across humanity was fairly even. For instance, estimates by the French statistician and demographer Paul Bairoch showed that China and India, which housed close to half of the world's population in 1750, accounted for 57% of the world's GDP.[4] This changed rapidly over the next 100 years. By 1830, India's share of world GDP had dropped from 24.5% to 17.6%, and by 1913 it fell even more precipitately to just 1.4%. China held its share at over 30% up to 1830, but then its share collapsed to 3.6% over the next 80 years (Figure 2.4).

Several factors contributed to this rapid and accelerating divergence in the 19th and 20th centuries. Among these were Europe's discovery and exploitation of the resources of North and South America and its

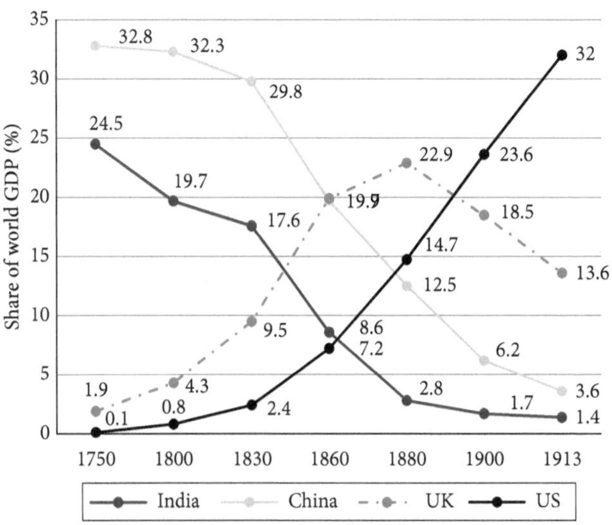

Figure 2.4 Results from Paul Bairoch's study of the world economy.
Source: Created by the authors based on Bairoch, Paul, *Economic and World History: Myths and Paradoxes*, University of Chicago Press, 1993.

simultaneous colonization of large parts of Asia, notably India, Burma, Indonesia, and southern China. The engine that drove both was the harnessing of fossil fuels for purposes of industry and transport.

Fossil fuels and industrialization

Britain began the shift into coal in the 18th century and became the world's first hegemonic power in the 19th. The US began the shift into coal and oil in the mid-19th century and succeeded Britain as the world's second hegemonic power in the second half of the 20th century Japan began its industrialization in the 1880s and 1890s and became a third centre of industrial power in the 1920s and 1930s.

In the second half of the 20th century, a few of the smaller countries in East Asia—notably Taiwan, South Korea, Malaysia, Singapore, and Thailand—took advantage of *les trente glorieuse*, also described as the Golden Age of Capitalism, the 30-year wave of prosperity that swept over the world after the end of the Second World War. This led to a growing disparity in wage rates between the US and Europe, on the one hand, and East Asia, on the other. The surge of demand created by the Vietnam War contributed to the development of an export-led growth model which propelled the industrialization of East Asia societies to close to US and European levels by the end of the 20th century.

But by the 1980s, this final surge of growth among late starters had virtually exhausted the space in the global market for further industrialization. In a prophetic paper on the subject, written in the early 1990s, Harvard economist Dani Rodrik (identified in private communication) pointed out that in all but four or five of the countries that had begun their drive to industrialize after the end of the Second World War, the share of industry in their GDP had stalled at 18–25%. The lone exception was China, which broke the economic chains of a statist, planned economy in 1981–83, made a Herculean effort to attract foreign investment and acquire state-of-the-art technology, and achieved an unprecedented rate of industrialization that no market economy could emulate.

India missed the first industrialisation bus, but could have caught the second, in which Japan and Italy found a place because the Mughal Empire had already created two of the necessary preconditions for the rise of capitalism in India. These were the creation of a single market across the entire subcontinent, and a network of trade-cum-banking houses whose *hundis* (promissory notes) were respected in far-flung parts of the country. These developments were contemporaneous with the development of the market fairs in Europe, called the 'Piacenza fairs', that began the knitting of Europe into a single market. In Europe, this was made possible by the spread of a Genoa-based banking network that honoured each other's promissory notes and made it therefore unnecessary for merchants to carry gold or silver to finance their purchases.

In India, the *hundi* system performed the same function. In fact, the willingness of banking houses such as the fabulously wealthy House of Jagat Sheth in Bengal to issue *hundis* in favour of the British East India Company made it possible for 'the Company' to wage war against Indian rulers in distant places without having to carry mountains of rations or lay waste to the countryside through which its armies passed in order to feed its army.

The third essential component of a modern market economy, a safe and rapid transport system, was created by the British. This was a network of railways that facilitated the rapid movement of goods from their points of production to the market—in this case to the ports of export. The first 34-km railway line from Thane, in the interior of Maharashtra, to the port of Bombay was completed in 1853. This was 19 years before Japan built its first railway line covering a similar distance, from Tokyo to Yokohama.

By the end of the 19th century, the British had built networks of railways around all three of India's major ports, Calcutta (now Kolkata), Bombay (now Mumbai), and Madras (now Chennai), and transcontinental lines that linked Punjab in the far north to both Calcutta and Bombay. This was considerably ahead of railway development in Japan.

India was better endowed for industrialization than Japan in one other respect: it had vast reserves of coal. By contrast, in relation to its needs, Japan had very little coal[5] and had to invade Korea in 1910 to gain access to the coal deposits in North Korea to meet its growing need for energy. India, by contrast, had abundant coal reserves in three areas of the country: Central India, Assam, and the lower half of the Gangetic Valley. Bihar in particular had large reserves of anthracite coal, with up to 90% carbon content. The British used it to run their locomotives and eventually all but exhausted its seams, but not before Jamshed Tata, the first truly great modern entrepreneur of India, built the first steel plant in Bihar to take advantage of this abundance of coking coal.

Why then did Japan forge ahead in industrialization and India get left far behind? The reason was that Japan was an independent country and India was a colony of the British. So, while Japan was able to close its frontiers to foreign manufacturers to develop its indigenous industry, Indian entrepreneurs wishing to do so were prevented by ingenious obstacles too numerous to relate. In spite of this, Tata built his steel plant in 1905 in the teeth of British duty-free imports. It took 16 years for Tata to pay the first dividend to its shareholders. Tata went on to make the Tata Iron and Steel Company one of the most efficient steel plants in the world by 1951, as rated by the World Bank.

Trade with Britain, monopolized by the East India Company, also destroyed Bengal's thriving handicraft-based cotton textiles industry when the company began to import cheap mill-made cloth from Manchester in 1813. The effect on rural employment and incomes was disastrous.

Amiya Bagchi and other economic historians have estimated that the share of industry in the labour force fell from 21.6% in 1809–13 to 8.5% in 1901. Textiles were the hardest hit: the share of spinners and weavers fell from 12.6% to 1.5%.[6]

But the absence of protective tariffs could not prevent a modern cotton textiles industry from springing up between 1870 and 1890 in western India around Ahmedabad and Bombay. In 1947, when it gained its freedom, India was one of the leading textile exporters in the world. Indian firms had also established a significant presence in 16 other sectors of industry, ranging from steel and engineering to jute. All of these had developed in the teeth of a zero-import duty regime imposed by the British through a doctrine of Imperial Preference which allowed the Government of India to impose tariffs on imports from every country in the world except Britain.

Having failed to persuade London to allow protective duties to be imposed on manufactured imports, Indian industrialists began to finance the freedom movement. All it asked for in return was that the Congress Party press Britain to impose protective tariffs on manufactured imports. The Congress did press the British government to do this, but to no avail.

When India gained its freedom, industry fully expected to receive the protection for which it had been asking. But to its immense disappointment, enamoured by the Soviet success in mobilizing science, manpower, and materials during the Second World War, the Congress Party opted for a 'socialist', state-owned model of planned development instead of the capitalist alternative. Instead of promoting an already highly efficient private sector, the government enacted a spate of anti-private-sector laws that not only stifled the growth of Indian companies but also drove already established foreign-owned enterprises out of the Indian market. By doing so it cut itself off not only from foreign capital, but also from access to the technology revolution that was transforming European and American industry in those first 30 years after the end of the Second World War.

Responsibility for industrializing India fell therefore entirely on the shoulders of the state-owned enterprises set up by the central and state governments that came collectively to be known as the public sector. To minimize dependence on imports, the government adopted a heavy investment-led growth strategy. 'We will save foreign exchange by making the machines that make the machines that meet our needs' was the slogan the government adopted. This strategy proved a disastrous failure. On the one hand, it resulted in one of the slowest rates of GDP and industry growth among the developing countries—3.6% and 4.8% respectively—for a quarter of a century (1956 until 1981). On the other hand, it created very few jobs in the non-agricultural sector of the economy.

The tentative beginning of liberalization

India began to experiment gingerly with economic liberalization at the end of the 1970s. Imports were liberalized in 1978 and many of the crippling constraints that had been imposed on private investment in the 1960s were removed in the mid-1980s. This led to a burst of industrial growth in the second half of the 1980s, with GDP growth exceeding 6% per year during the entire period.

But the reforms were incomplete. Imports were severely controlled. So the rupee remained overvalued. This made it difficult to increase exports, and especially difficult to develop new products for the world market. That had been China's experience also up to the early 1980s and was the single most important reason for Deng Xiaping's dramatic opening up of China's domestic economy to international investment in the early 1980s. But India's very different politics ruled out any such dramatic shift. So, throughout the 1980s, economic liberalization had to be carried out piecemeal and earned the derogatory sobriquet of 'Reforms by Stealth'.

As a result, exports did not grow, and the increased demand for foreign exchange needed to fuel the higher rate of growth had to be met through loans from the overseas capital market. As the international banks grew more anxious about India's capacity to repay, these loans had to be taken increasingly from the short-term money market.

A foreign exchange crisis was bound to take place sooner or later, but it was brought on with great suddenness by Iraq's invasion of Kuwait in August 1990. Fresh lending dried up quickly, while repayments continued to climb week after week. To stave off bankruptcy, the Reserve Bank of India, India's central bank, had to pawn the 55 tonnes of gold it held in its reserves. The shock this gave to Indian public opinion destroyed the last resistance of the socialists to the switch to a market economy. In the next 2 years the rupee was devalued to a third of its pre-liberalization value against the dollar; all controls on imports and exports were lifted; and all the remaining anti-monopoly and anti-foreign investment laws were scrapped.

The economy responded immediately. Between 1993 and 1997 GDP growth averaged 7.2% and industry grew faster. After a shallow recession in the next 5 years, when it dropped to 5%, GDP growth took off again in 2003 and averaged close to 8% up to 2010, with industry recording even higher rates of growth.[7] But unlike China, India was unable to sustain this growth beyond 2010. In the next 10 years the growth of industry virtually stalled. In December 2021, industrial production was only 38% higher than it had been in 2011–12. This amounted to an average growth of just a little over 3% a year.[8]

This collapse of industrial growth brought the growth of GDP down to 4.04% in 2019–20[9] (Figure 2.5). Most of this drop took place over nine quarters, from January 2018 to March 2020. This was before COVID-19 forced a lockdown on the nation at the end of March 2020.

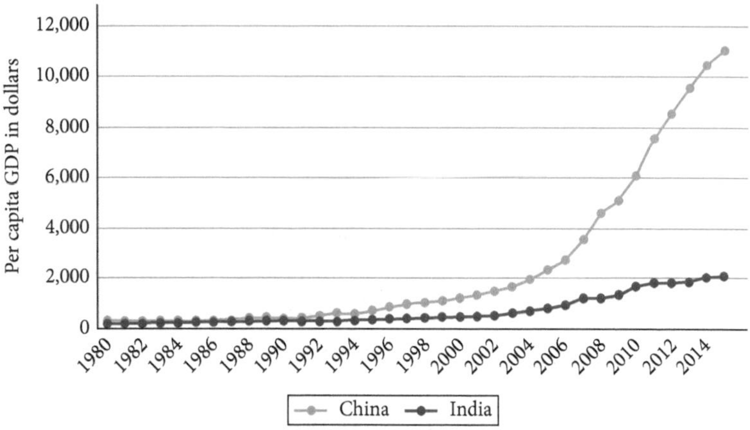

Figure 2.5 China vs. India: nominal GDP.

Seventy years after the framing of India's first Five Year Plan, after the failure of communism, the dissolution of the Warsaw Pact, the break-up of the Soviet Union, and the phenomenal success of China and the Association of Southeast Asian Nations (ASEAN) countries, which had opened their gates wide to foreign investment in the 1970s and 1980s while India kept its doors tightly shut, the folly of India's statist policies is something no one in India likes to talk about. In 1981, India's per capita income was marginally higher than China's. This remained true up to 1990 when China's per capita GDP was $318 and India's $368. Then the Chinese economy took off: by 2021 its per capita income had risen to $11,891, while India's remained at $2,257[10] (Figure 2.6).

Most of the ASEAN countries have also left India far behind. In 2021, Malaysia's per capita income was $11,125, Thailand's was $7,809, and Indonesia's was $4,225. These numbers were far behind the achievements of the original 'Little Tigers' of Asia: Taiwan's per capita income had risen to $33,409 and South Korea's to $35,196. Meanwhile, India lagged with a per capita income of $2,116.[11] This is what an 'average' European or American spends in 2 weeks.

The most disquieting feature of India's growth experience is that even the high rate of GDP growth that it enjoyed from 2003 to 2012 has proved unsustainable. That failure has further exacerbated two crises that have been deepening over the past 30 years or so and have reached a point

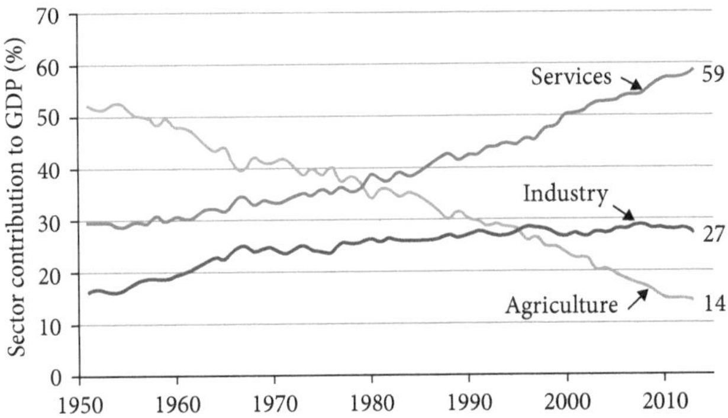

Figure 2.6 Sector share of total GDP, 1951–2013.
Source: Wikimedia Commons, 1951 to 2013 Trend Chart of Sector Share of Total GDP for each year, India.png. https://commons.wikimedia.org/wiki/File:1951_to_2013_Trend_Chart_of_Sector_Share_of_Total_GDP_for_each_year,_India.png (viewed 20/1/2022) (https://creativecommons.org/licenses/by-sa/4.0/deed.en).

where they can no longer be ignored. These involve an acute crisis of unemployment and an equally intractable crisis in agriculture.

These dwarf two other, older crises from which India has been suffering. The first is a perennial shortage of foreign exchange that has brought on two economic crises, in 1966 and 1991, and continues to dog the country today. The second is a crisis of governance that has led to a chronic neglect of the elements of public policy that enable rulers to pass on the benefits of economic growth to the ruled. These are the provision of infrastructure, health and education services, the elimination of air and water pollution, and safe treatment of municipal waste.

With so much time lost and so much leeway to make up, India cannot treat the shift out of fossil fuels as a challenge that is distinct from all the others that it faces. It simply has to find ways to avert the threat from climate change that also make it possible for it to meet the challenge for survival and sustainable growth.

Notes and References

1. In February 1998, to commemorate 50 years of independence, the Archaeological Survey of Pakistan organized a vast exhibition of the Indus Valley civilization at the Asia Society in New York in which it displayed early examples of Indus Valley pottery that were carbon dated to 3500 BCE.

2. Archaeological Survey of Pakistan, notes for an exhibition organized at the Asia Society in New York in February 1998 to celebrate the 50th anniversary of Pakistan's independence.
3. Ibid.
4. Bairoch, Paul (1993). *Economic and World History: Myths and Paradoxes.* University of Chicago Press.
5. In 2016 Japan had 386 million tonnes (Mt) of proven coal reserves left, 0.0003% of world reserves. The main reason why it annexed Korea in August 2010 was to gain access to North Korea's coal deposits.
6. Bagchi, Amiya Kumar (1976). Deindustrialization in India in the nineteenth century: some theoretical implications. *Journal of Development Studies* 12(2):135–64. Quoted in Clingingsmith, David and Williamson, Jeffrey G. (2005). *India's Deindustrialization in the 18th and 19th Centuries.* Harvard University.
7. StatisticsTimes.com: GDP Growth of India 2021, source from World Bank. https://statisticstimes.com/economy/country/india-gdp-growth.php (viewed 18/1/22).
8. Ministry of Statistics and Programme Implementation: Quick estimates of industrial production for December 2021. https://pib.gov.in/PressReleseDetailm.aspx?PRID = 1797666 (viewed 19/1/22).
9. Ibid.
10. Vivek Kaul: The tale of two economies: what changed in 30 years. https://www.livemint.com/news/india/the-tale-of-two-economies-what-changed-in-30-years-11592930762996.html. See also IMF World Economic Outlook, October 2021.
11. Ibid.

3
The Crisis of Unemployment that India Cannot Evade

In the preceding chapter we described why India, which could easily have begun its industrialization shortly after Britain, missed both the first wave of industrialization that swept through Europe in the 19th century and the second wave that spread to Asia in the 20th century. We turn now to the constraints within which India is attempting to complete this transition today and the initiatives it will need to take to overcome them. Chief among the challenges it faces is the need to do this while simultaneously reducing, and eventually eliminating, its reliance on fossil fuels.

As we pointed out in Chapter 1, this challenge, which requires developing countries to decarbonize their energy consumption without sacrificing economic growth, is far more daunting for the low- and middle-income countries of the 'South' than it is for the industrialized countries of the 'North'. In its report titled 'World Energy Investment 2021' the International Energy Agency (IEA) admitted this bluntly when it stated: 'The gap between today's investment trends and a sustainable pathway is larger in emerging market and developing economies . . . in economies where governments have more fiscal space and are able to borrow at low rates, recovery strategies offer a major opportunity to boost investment in infrastructure, efficiency and clean energy technologies'.[1] 'Fiscal space' and low interest rates are precisely what the developing economies do not have. India is no exception.

The average saving/investment rate in the EU has been 12% during the past more than two decades. The average savings rate in the US between 2000 and 2020 before the COVID-19 pandemic hit averaged 17%. In both regions a sizable part of this saving was already being invested in the energy sector. Between 2000 and 2019, the EU spent between 4 and 5% of its GDP on meeting its energy needs, and the US, which saw

Sun, Wind, and Biomass. Prem Shankar Jha and Michael B. McElroy, Oxford University Press.
 DOI: 10.1093/9780198944768.003.0003

the development of fracking, spent close to 6%.[2] So the task before their governments in addressing the climate challenge is essentially to persuade their energy and allied industries to phase out investment in fossil fuels and shift progressively into renewable energy. As the IEA's 'World Investment Outlook 2021' pointed out, the additional sums needed to speed up this shift could easily be raised in an international capital market that was flush with funds and starved of new investment opportunities, a condition reflected by the historically low interest rates it was offering on fresh deposits.

Developing countries face a more difficult task—India's twin crises of growth and employment

The task that the developing countries face is more daunting. Since 1993, when it abandoned its plans for a market economy, India's GDP has grown at an average annual rate of 6.25%.[3] As these decades included two periods of economic boom, from 1993 to 1997 and from 2003 to 2011, interrupted by mild recessions from 1997 to 2002 and from 2007 to 2009, with only a gradual decline in growth from 2016 to March 2020, it would be safe to assume that Indians have grown used to a 6%-plus annual rate of GDP growth and will not easily accept the hiatus in the improvement of living standards that a change in the goals of investment from development to decarbonization would bring. This is because even after more than three decades since 1993 of relatively high GDP growth, the country continues with every passing year to suffer more grievously from poverty, inequality, unemployment, air and water pollution, and living conditions that have become more and more difficult especially in the cities.

What are the options for India to maintain its projected economic growth while at the same time facing up to its responsibilities in terms of climate change? There are only two: reduce the energy content of GDP growth, or reduce the carbon content of energy consumed. Over the past two decades the reduction in the former has been significant. The Energy and Resources Institute (TERI) has estimated in a study that the energy required per unit of GDP declined by 1.58% a year between 2006 and 2021, and that even in a 'business-as-usual' scenario this improvement is likely to be maintained and could amount to 1.61% a year between 2006 and 2031.[4] A further improvement in energy efficiency beyond this is unlikely since most of the improvement since 2006 has resulted from a sharp slowdown in the growth of manufacturing and a rise in the share of the services sector in the GDP. This cannot be sustained for much longer.

The share of renewables in India's power generation capacity has risen from 15.22% in 2010 to 20.08% in 2021, recording an annual growth of 2.55%.[5] Since electricity accounts for only a third of the energy consumption of the country, this amounts to a spontaneous annual rate of shift out of fossil fuels of 0.85%. Fuel-efficient motor car engine designs and the continued electrification of the railways have also increased the energy efficiency of the transport sector. This will improve further as the government's programme for increasing the ethanol content of gasoline and the electrification of two wheelers and three-wheeler rickshaws gathers steam. However, even under the most optimistic projections, it is unlikely that even the sum of these two developments can succeed in decreasing the carbon content of energy consumption by more than 1.5% a year.

India is not a lone example. It is doubtful whether any developing country in the world can finance such a rapid rate of shift out of fossil fuels from domestic resources alone. India's savings rate had previously reached a peak of 36.9% in 2011–12 at the height of its 8-year-long boom, before sliding with slower growth to 28.2% in 2020–21 (Figure 3.1). The former saving rate tempts us to conclude that if the present rate could be

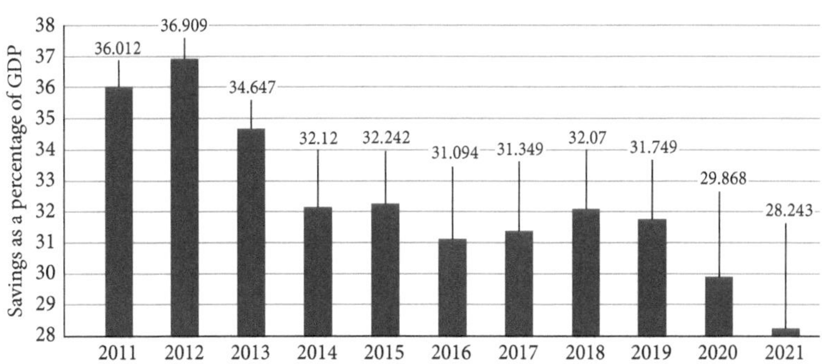

Figure 3.1 Savings as a percentage of GDP.

Source: Created by the authors based on Economic Indicators, CEIC, India Gross Savings Rate (© 2021 CEIC Data, an ISI Emerging Markets Group Company). https://www.ceicdata.com/en/indicator/india/gross-savings-rate (viewed 4/2/2022).

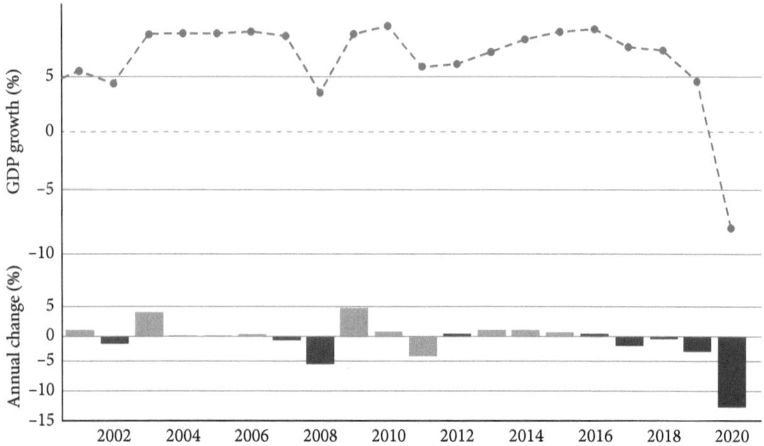

Figure 3.2 GDP growth rate 2000–20 and the annual change in it.
Source: World Bank Data, MacroTrends, India GDP Growth Rate 1961–2024. https://www.macrotrends.net/countries/IND/india/gdp-growth-rate (viewed 4/2/2022).

ramped back to 37%, India might be able to finance both its continued growth and most of its shift out of fossil fuels from its own resources. But this would be wishful thinking, because the high saving rate attained in 2011–12 was the product of unprecedented growth ranging from 7.86% to 8.50% per year over the previous 8 years. When the growth rate declined, it too fell. It will rise again only if the rate of GDP growth also reverts to its pre-2011 level (Figure 3.2). But that will not happen if a sharply increased proportion of the savings is diverted to replacing one source of energy with another, for it is possible that many more people could be thrown out of work in the industries and coal mines that will lose orders or be closed down than will obtain employment in the highly capital intensive new power generation companies and their supply chains. The loss of jobs, and therefore of income and savings, will be even greater if the import content of the new energy enterprises is much higher than the import content of the enterprises they are replacing.

The composition of savings in developing countries provides a second reason to believe that even a 37% rate of saving might not be able to finance both decarbonization and growth at the same time. This is the fact that not all the savings in them takes place in financial form and is therefore not fungible. In India, approximately one-third of national savings

Table 3.1 Sector-wise savings (in Rs trillion and current prices), 18 September 2020.

Year	2015–16	2016–17	2017–18	2018–19
Gross savings (of which):	42.82	48.25	55.38	57.13
Deduct household savings in physical assets	13.6	16.4	19.5	22.1
Equals financial saving	31.2	31,8	35.9	33.1
Financial saving/GDP (%)	23	21	20.8	18

Source: Compiled from Reserve Bank of India (RBI) Publications. Table 11: Sector-wise domestic savings (at current prices). https://www.rbi.org.in/scripts/PublicationsView.aspx?Id=19744 (viewed 2/9/2022).

is deployed through the direct creation of physical assets: the repair or purchase of a house, a shed, a tractor, or a tube well; the purchase, levelling, and irrigation of a tract of land; the construction of a fence; and other such improvements. A good part is also invested in the purchase of gold and silver jewellery and ornaments, or even in buffaloes and milch cows, because gold, silver, and animals are assets that can be sold at very short notice, should an urgent need for money arise. Only two-thirds of gross national savings takes the form of financial assets such as bank deposits, shares and debentures, pension funds, health insurance, and retirement schemes. These are the only capital assets that can be redeployed to achieve specific national goals. The 11.5% annual growth of investment in renewable energy can therefore only come out of this portion of national savings. Table 3.1, constructed from data assembled by the National Statistical Office and the Reserve Bank of India (RBI), gives a precise estimate of the financial savings that are available for investment in the monetized sectors of the economy.

India's crisis of unemployment

India is facing a crisis of unemployment that has grown deeper by the day, and no government has so far been able to resolve it. Sixty-five percent of its people still live in villages and adjoining small market towns, and half of them continue to depend for their livelihoods on agriculture. In the six decades up to 2011, a 2% growth of population on the one hand and slow growth of industry on the other have steadily increased the pressure on the land. Where the average ownership holding of land was 2 hectares in 1970–71, it is 1 hectare today. A vast, young, mostly rural population is therefore in urgent need of an assured source of income and a secure future. And no one, within or outside the Indian government, has the faintest idea of how to provide this.

The only panacea that policy makers and economists have been able to think of so far is faster industrial growth. For a short while this seemed to be working: between 2003 and 2011, the economy averaged close to 8.5% GDP growth, and created 7 million jobs annually in industry and the services sectors. Three million of these jobs were taken by urban youth, but the remaining 4 million were filled by rural youth drawn away from agriculture and its ancillary occupations.[6] Industrial growth has declined from over 9% during the boom years to recently below 4%. The share of manufacturing in GDP has stalled at 18%, and more and more of this manufacture is being automated, computerized, and carried out by robots. According to the Government of India's Periodic Survey of Employment 2017–18, more than 100 million persons between the ages of 15 and 64—a quarter of the entire working population—are not merely without any regular employment, but have given up looking for it and are surviving on seasonal and casual work. Despite this there is a curious reluctance among economic forecasters to acknowledge this terrifying backlog of young people without work. A McKinsey report published in August 2020[7] summed up the challenge India faces as follows:

> India's GDP will need to grow 8 to 8.5 per cent annually over the next decade, nearly double the 4.2 per cent growth rate registered in the financial year 2019–20, to create 90 million new non-farm jobs between 2023 and 2030. Of these 60 million will be new workers who will enter the workforce; an additional 30 million workers need to be moved from farm work to more productive non-farm sectors.

Ninety million persons is more than the entire population of Germany, the largest country in the EU, and almost one-and-a-half times the population of France. But it is only 2 million more jobs per year than were created in India during the boom years of the 2000s.

For India to achieve an 8–8.5% average rate of GDP growth between 2020 and 2030, as currently projected, seems unlikely from the vantage point of the end of 2024. To achieve this by merely doing what it is already doing but only better, as the McKinsey report suggests, is whistling in the dark.

What the McKinsey report is silent on is the huge backlog of unemployment that has built up in the country. This has been the product of slow and capital-intensive economic growth on the one hand and a simultaneous explosion of population on the other. Between 1951 and 2001, India's GDP averaged growth per year of a shade over 4%. But in precisely the same period it experienced a sharp bulge in the rate of growth of its population, from just under 1% between 1921 and 1931, 1.4% between 1931 and 1941, to just under 2.4% a year between 1951 and 1991. In the first two decades of the 21st century, annual population growth slowed, first to 1.77% in 2001–2011 and then to 0.95% in 2011–2021. This slowdown reflects the spread of access to birth control and an increasing desperation among the poor to ensure a reasonable future for their children.[8]

Unfortunately, it has been paralleled by changes in the rate of growth of GDP, which, after reaching an explosive 8.5% in 2010, fell gradually to 3.74% in 2019–20 before the onset of the COVID-19 pandemic. The population bulge seems destined to continue subsiding, but it has left the country with huge numbers of young people in search of work for whom there are not only no jobs but also dimming chances of even obtaining casual employment that can suffice to keep body and soul together.

Today, unemployment and huge underemployment are by far the most serious economic problems the country faces. The main cause is the slow and capital-intensive growth imposed on the country by the centralized, autarchic model of growth that prevailed from 1951 to 1991. During the first three of these four decades, when the Planned Economy was at its height, growth averaged 3.6% annually. After liberalization 'by stealth' began in 1978, it rose to 5.4% in the 1980s. For four whole decades, therefore, while East and South-East Asia raced ahead, India's growth averaged just over 4%.[9]

Unemployment was not a significant issue for the working age population in pre- independence India because the population of the whole of British India, i.e. India, Pakistan, and Bangladesh, was only 388 million and had grown at a snail's pace from 279 million 50 years earlier. There was therefore abundant land for extending agriculture, with British Army recruitment during the two World Wars as an option for fallback.[10] But those days have long passed. Agriculture continued to absorb new income earners by bringing cultivable areas under the plough. Much of this was done by cutting down huge swathes of virgin forest in the Terai, the verdant foothills of the Himalayas. This continued until 1960–61 when nearly all of this land was exhausted. The central and state governments then began to create jobs within the public sector to absorb the jobseekers. As a result, total employment in the two levels of government increased from 7 million in 1961 to 19 million 30 years later, on the eve of the economic crisis in 1991. Another 2 million were absorbed into an increase in the armed forces after the Sino-Indian border war of 1962 and into state armed police cadres needed to deal with rising social unrest. But the reforms of 1991–92, which required a much higher level of fiscal discipline within the state apparatus, closed this route. An idea of the impact this had was given by a report of the Planning Commission Task Force on Employment Opportunities in 2001. The report began by pointing out that employment growth had dropped from 2% between 1983–84 and 1993 to less than 1% over the next 6 years. This meant that, prior to the economic reforms, more than half of the employment had, for years, been created by the government itself.[11] Economic reforms laid bare the crisis that the government had been ignoring.

Agriculture's potential for creating employment was exhausted. Every square metre of arable land was being cultivated. The pressure of population on the land was growing fiercer by the day. Land holdings had been subdivided generation after generation until the average holding had fallen from 2 hectares in 1970–71 per family in 2001 to 1 hectare per family in 2021. As many as two-fifths of these holdings could no longer feed their families, so their owners were sending one or more family members to work in the cities.

Instead of tackling the emerging problem head-on, government after government in New Delhi chose to look the other way. In this, its behaviour was sharply at odds with that of Taiwan, South Korea, and China, where economic liberalization policies were driven, from the start, by the need to provide employment for their citizens. The Indian government chose to hide its failure behind changes in the definition of employment. It had, from the very beginning, defined employment not as *jobs* earning salaries, but as *work* that earned livelihoods. Furthermore, on the plea that data on livelihoods are far more nebulous and therefore harder to collect than data on employment, it had tasked its National Sample Survey Organization (NSSO) with doing this every 5 years instead of annually. This led to a format of reporting that, for laymen, obscured, instead of highlighting, the state of unemployment in the country.

Recognizing that large numbers of people have only seasonal work, the NSSO initially decided to collect two kinds of data to estimate employment in the unorganized sector of the economy. It called these *usual* status and *weekly* status employment. But as the employment crisis deepened, it found that sizable numbers of workers were working for only a few days, or even only a day at a time. So, it created a third category of employment and called it 'Daily Status' workers. From then on, one had to have worked for only a few days every month to be counted among the employed. And, one had to have been looking for work and not have found it for even a few days in the week or month to be counted as unemployed. If you had looked for work, failed to find it, and had given up looking *you ceased to be counted as unemployed.*

To tally the number of unemployed, the NSSO defined three states of activity for all those who belong to the labour force (i.e. all those between 16 and 60). These are: (1) Working, i.e. engaged in an economic activity, no matter for how long. These are considered employed; (2) Seeking or available for work during the reference period of the survey. These are considered 'unemployed'; and (3) Neither seeking nor available for work. These are no longer counted as unemployed.[12] The only way to estimate the true state of employment is to examine the 'labour force participation rate'. The fall in this parameter within a given period, if any, tells us how many persons have ceased, whether voluntarily or involuntarily, to look for work. According to the government's Ministry of Statistics and Programme Implementation, the rate had reached an all-time high of 63.7% in 2006–07, 2 years into the 9-year boom that began in 2003–04, and had then fallen, at first to 57.1% in 2011–12 and then more rapidly to 49.8% in 2019, the year before the onslaught of COVID-19 (Figure 3.3).

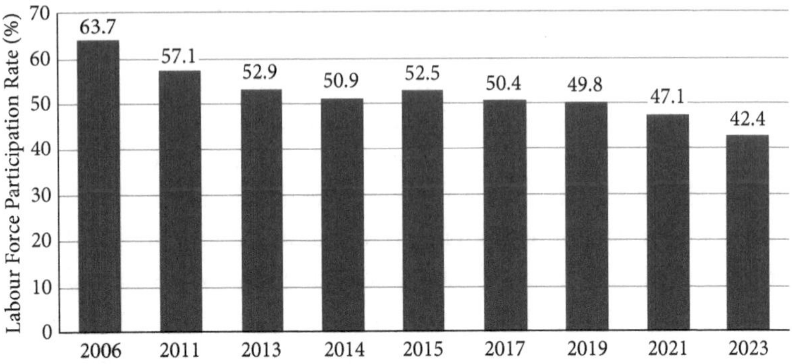

Figure 3.3 Labour force participation rate.
Source: Created by the authors based on data from Ministry of Statistics and Programme Implementation, Government of India.

The decline between 2004–05 and 2011–12 was caused most likely by the withdrawal of women and young adults from the rural workforce transitioning to minding their own tracks of land and performing essential services as more men were drawn to non-agricultural jobs in the cities. But the decline from 2012 to 2019 was almost entirely the result of a slump in urban employment growth and a return of the men to agriculture.[13]

India's labour force, as defined by the NSSO, was 446.1 million in 2005, 457.5 million in 2011, 482.7 million in 2019, and 501 million in 2021. Thus, while 18 million people between 16 and 60 left the workforce during the boom years, almost certainly voluntarily, another 21 million did so almost certainly involuntarily, in the 8 years that followed. The savageness of the reversal is revealed by an annual sample survey of employment, called the Periodic Labour Force Study (PLFS), of the National Sample Survey (NSS). This showed that for both males and females, urban unemployment among those with a secondary or higher education fell between 2004–05 and 2011–12, but then rose at a dizzy pace over the next 6 years. This reversal is shown in Tables 3.2 and 3.3.

The tables confirm that women were the first to lose their jobs in both categories of employment when the boom years ended. A more detailed analysis of the PLFS data by University of Bath professor Santosh Mehrotra showed that among the youth who lost their jobs, half were

Table 3.2 Urban educated unemployment rates, 2004–05 to 2017–18 (percentages of the total labour force, which is defined as all those between 16 and 60).

Year	2004–05	2011–12	2017–18
Female	15.6	10.3	19.3
Male	5.1	4.0	9.2

Source: Periodic Labour Force Study of the National Sample Survey (various years).

Table 3.3 Youth unemployment rates, 2004–05 to 2017–18 (percentages of the total labour force, which is defined as all those between 16 and 60).

Year	2004–5	2011–12	2017–18
Rural male	3.9	5.0	17.4
Urban male	7.5	8.8	18.7
Rural female	13.1	14.9	27.2

Source: Periodic Labour Force Study of the National Sample Survey (various years).

below 26 years in age and two-thirds were below 35. What is more, job loss not only was greater among the educated than the uneducated but also *increased with the level of education.* While unemployment among those who had completed their secondary education, i.e. completed studies up to class 10, rose from 4% to 16% over these 7 years, unemployment among those who had completed their higher secondary studies (class 12) rose from 7% to 22%, among those with Bachelor's degrees from 20% to 38%, and among those with Master's degrees from 19.9% to 43%. The job loss among those with technical degrees was similar: from 19% to 38.6%.[14] These data are disturbing because they show that education is no longer a way out of poverty. The growing despair of the youth may be judged from the fact that in February 2018, when the Indian Railways announced that it would be filling 63,000 vacancies for class III and class IV (clerical and menial) jobs, they received *19.1 million applications.*[15] A large proportion of the applicants had Bachelor's and Master's degrees. In 2021, another announcement of 35,000 clerical jobs drew

12 million applicants. When the Railways, in desperation, tried to weed out a large proportion by raising the minimum qualification required from a high school certificate to a Bachelor's degree, the applicants, all of them young, rioted and burned down an empty train in Gaya, one of the largest cities in the state of Bihar.[16]

For the vast majority of youth, who now regularly fail to find even marginally secure and paying jobs, there is no option left but to migrate from the rural areas to the cities and join the crowds of job seekers who assemble every morning in the labour *haats* (informal marketplaces) hoping that they will be among the lucky ones that day. In 2020 on the eve of the COVID-19 pandemic, urban India hosted around 139 million migrant workers. Of this number, 50–60 million worked outside their home states. Other than a 'ration card' that entitles them to purchase subsidized food from government-approved shops, and the beginnings of a national health insurance system, they are entirely unprotected from the vagaries of the market and the hammer blows of fate.

The challenge associated with the requirement for India to transition to a zero carbon energy emission scenario is discussed in subsequent chapters. Chapter 11 explores the possibility of this transition taking place by 2050. The background assumption is that the GDP will grow over the intervening years from a level of $2.9 trillion in 2020 to $16.0 trillion in 2050. Final energy consumption is assumed to increase by a factor of 2.24 over this period. Chapter 11 assumes that energy in the zero fossil fuel 2050 case is supplied primarily by a combination of wind and solar with important contributions from hydrogen produced from electrolysis and gasification of biomass. An important conclusion of the analysis is that it would actually be less expensive to meet the final energy demand projected for 2050 using these renewable energy sources rather than relying on fossil sources as would be the case in a 'business-as-usual' scenario. Chapter 12 extends the analysis to consider additional sources of energy produced from gasification of biomass, sources that again would be carbon free in the climate context in that the carbon dioxide emitted from these sources would have been derived ultimately from the atmosphere. Chapters 14 and 15 discuss how renewable energy sources such as wind, solar, and biomass could be exploited to increase the opportunities for job creation specifically in poor rural areas, highlighting investments that could provide a market for agricultural waste that could be gasified to produce a variety of useful products, including electricity to fuel cold storage facilities that could enhance the income farmers could realize from the sale of their products.

Notes and References

1. IEA: World Energy Investment 2021, Executive summary.
2. BlackRock's (BLK) weekly market commentary, 6 April 2022. https://seekingalpha.com/news/3821184-blackrock-says-eu-spending-91-of-gdp-on-energy-more-us-fossil-fuels-needed (viewed 2/2/2022).
3. Calculated from World Bank data on India's GDP growth rate. https://www.macrotrends.net/countries/IND/india/gdp-growth-rate (viewed 2/2/2022).
4. Mathur, Ritu, Sethi, Girish, and Kumar, Sachin (2018). Energy efficiency & India: the possibilities ahead, TERI, 29 May 2018. https://www.teriin.org/article/energy-efficiency-india-possibilities-ahead.
5. Statista. https://www.statista.com/statistics/1300330/india-renewables-share-electricity-mix/ (viewed 4/2/2022).
6. Interview of Santosh Mehrotra by Mitali Bannerjee of The Wire. https://www.youtube.com/watch?v=17kTGYT3CVY (viewed 2//9/2022). Mehrotra's research showed that after 2011–12, the number of people below the poverty line increased by 76 million, from 216 to 283 million.
7. McKinsey Insights: India's Economic Turning Point, 30 October 2020. https://www.mckinsey.com/in/our-insights# (viewed 9/2/2022).
8. https://www.macrotrends.net/countries/IND/india/population-growth-rate (viewed 2/9/2022).
9. Calculated from World Bank estimates of GDP from 1951–52 to 2013–14, with 2004–05 as the base year for indexation.
10. Census of India 1941, p. 62.
11. Planning Commission: Report of the Task Force on Employment Opportunities 2001. Para 1.5, p. 2.
12. For a detailed description of categories see 'Concepts and Definitions used in NSS'. https://www.mospi.gov.in/sites/default/files/publication_reports/concepts_golden.pdf.
13. https://tradingeconomics.com/india/labor-force-participation-rate#:~:text=Labor%20Force%20Participation%20Rate%20in%20India%20averaged%2050.13%20percent%20from,the%20second%20quarter%20of%202020 (viewed 11/2/2022).
14. Interview of Santosh Mehrotra by Mitali Bannerjee of The Wire. https://www.youtube.com/watch?v=17kTGYT3CVY (viewed 11/2/2022).

15. Joanna Slater, *The Washington Post*, 4 January 2019. https://www.washingtonpost.com/world/asia_pacific/indias-railroads-had-63000-job-openings-19-million-people-applied/2019/01/04/77a82e94-edb3-11e8-8b47-bd0975fd6199_story.html.
16. Raksha Kumar: This burning train is a symbol of the anger of India's out-of-work youth, 16 March 2022. https://www.npr.org/sections/goatsandsoda/2022/03/16/1084139074/this-burning-train-is-a-symbol-of-the-anger-of-indias-out-of-work-youth (viewed 12/2/2022).

4

The Crisis in Agriculture: a) In Dryland Farming

Any agriculture that depends solely on the rain is hazardous. But combine the hazards of erratic rains with growing population, the relentless subdivision of once-viable farms into smaller holdings, an absence of non-agricultural employment in the vicinity of their homes, with the ever present threat of a monsoon failure, and the plight of marginal farmers and unemployed youth in the rural areas ceases to be bearable. In this chapter we seek to outline the dimensions of the crisis that farmers in the dryland farming sector have been in the grip of for more than three decades. This is compelling more and more farmers to take the only way left that is open to get them out of their predicament. This is to commit suicide: 10,349 farmers committed suicide in 2018. Two years earlier, the figure had been 11,379. This was not a sudden development. As many as 296,438 farmers committed suicide in the two decades stretching from 1995 to 2014, an average of nearly 15,000 per year. In short, while urban India convinced itself that the country had solved the problem of food scarcity through the Green Revolution, one farmer was committing suicide in the villages every 35–45 minutes.

The vast majority of the suicides have taken place in the dryland farming areas of the country which stretch from its centre-west down to close to its southern tip. The causes of the rise in the suicide rate in the 1990s ranged from repeated crop failures because of drought or pest attacks to frantic risk-taking by increasingly desperate farmers plunging them into irredeemable debt when the gambles they took failed. In a 2008 study, Arvind Panagariya, Professor of Economics at Columbia University, concluded that 'farm-related reasons get cited only approximately 25% of the time as reasons for suicide' but 'studies do consistently

Sun, Wind, and Biomass. Prem Shankar Jha and Michael B. McElroy, Oxford University Press.
 DOI: 10.1093/9780198944768.003.0004

show a greater debt burden and greater reliance on informal sources of credit amongst farmers who commit suicide'.[1]

Two other studies carried out in Vidarbha, the eastern region of Maharashtra, one in 2009 and the other in 2012, came to similar conclusions. The 2009 study ranked indebtedness (87%) and deterioration in economic status (74%) as the main causes that drove farmers to suicide.[2] The 2012 study applied a Smith's Saliency method to rank the reasons cited by farming families that had lost a member to suicide. Debt, much of it brought on by crop failure, ranked first among a litany of causes that included alcohol addiction, low produce prices, stress caused by failure to meet family responsibilities, the increased cost of cultivation, and relentless pressure from private money lenders to whom tenant farmers had to go because they lacked the title to the land that would make them eligible for loans from the rural banks. These money lenders charged extortionate rates of interest and insisted upon repayment in spite of the failure of the crop.[3]

The region of Vidarbha accounted for more than a fifth of all of the suicides between 1995 and 2014; all but a small number of the remainder occurred in western Maharashtra, Odisha, Telangana, Andhra Pradesh, Madhya Pradesh, Gujarat, and Chhattisgarh. These seven states account for 71.8 of the 83.1 mha of dryland farming in the country. Maharashtra alone accounted for more than a fifth of all the suicides between 1995 and 2014.

This is the most tragic manifestation of a widespread and deepening crisis in dryland agriculture. A second, less noticed indicator is that while the *number* of family holdings in agriculture increased from 138.345 million in 2010–11 to 146.45 million in 2015–16, reflecting the continued subdivision of holdings that has taken place with each passing generation, for the very first time the total *area* under cultivation *declined* by 1.77 mha (1.11%) over the same period.[4]

This was a reversal of the trend witnessed since the first census of agriculture was taken in 1971. Up to 2011, farmers had brought more and more land under cultivation at the expense of permanent pastures and the few remaining forests that had not been sequestered by the forest departments of the state governments. This creeping expansion reflected the lack of opportunities for succeeding generations of rural youth to make a living from agriculture. The decline in cultivated area after 2011 showed that the size of a significant number of ownership holdings had fallen below the minimum needed to sustain a family no matter what it chose to grow on it. Combined with a dearth of non-agricultural employment opportunities in the villages or nearby small towns, this was forcing the smallest landholders to lease out their holdings to their neighbours or to other members of their joint families and to migrate to the cities in search of work that would earn the cash income that their families needed. By 2014–15 there were an estimated 160 million migrant workers working in the cities and industrial zones of India.[5]

Six of the seven states that account for the farmers' suicides fall within the area that contains the bulk of the dryland agriculture of the country.[6] In these states, except in river valleys and coastal deltas, the soil is poor and rainfall is scant. So crop failures are frequent and, as of now, there is very limited crop insurance available to the farmer. Farmers have been relying therefore more and more heavily on tube wells and open wells to reduce their vulnerability to drought. But the groundwater too is running out, as indicated by the hydrogeology maps displayed in Figures 4.1 and 4.2. Further, the areas of the country most dependent on groundwater for cultivation not only have the most meagre availability of this resource but also have exploited it to the point where further withdrawal has become unsustainable.[7] To describe the situation as a crisis would be an understatement.

Table 4.1 shows that as far back as 2006, groundwater extraction had reached, or exceeded, 100% of net annual recharge in one of five locations (1,065 out of 5,273) where assessments had been made. Five years later, in 2011, another more detailed assessment showed that while the proportion of aquifers in which annual water extraction equalled or exceeded annual recharge had remained more or less the same (1,284 out of 6,607 observations), 617 other aquifers had reached a 'sub-critical' stage, where the annual withdrawal was 70–90% of annual recharge. In addition, in 92 blocks (the fundamental unit of rural administration containing an average of 100 villages), the entire underlying aquifers had become saline. All in all, 31% of the groundwater aquifers were either overdrawn, dangerously close to it, or already saline.[8]

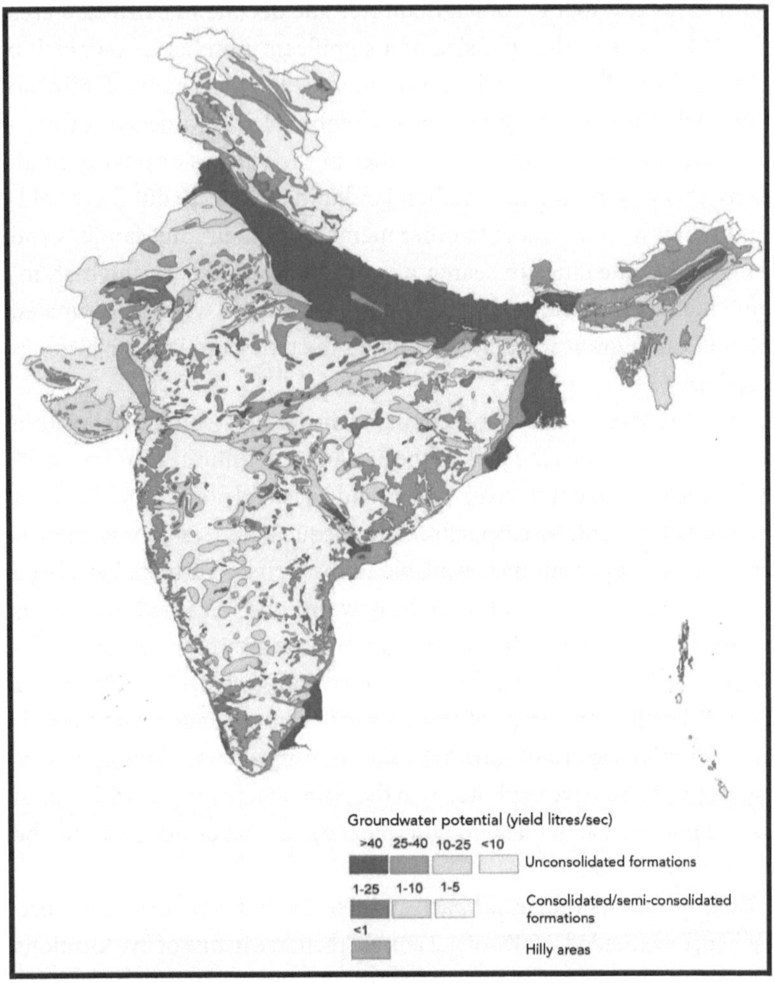

Figure 4.1 Map of groundwater availability, 2006.
Source: Towards Better Management of Ground Water Resources in India. B.M. Jha, Chairman, and S.K. Sinha, Chief Scientist of the Central Ground Water Board, Government of India.

Impact of global warming

Global warming has further aggravated the scarcity of water. A searing El Niño that propelled drought in 2016 was followed just 2 years later in 2018–19 by a second even more severe 10-month-long drought in central and southern India. So grave was the shortage of water that village children carrying empty jerry cans were boarding local trains to and from

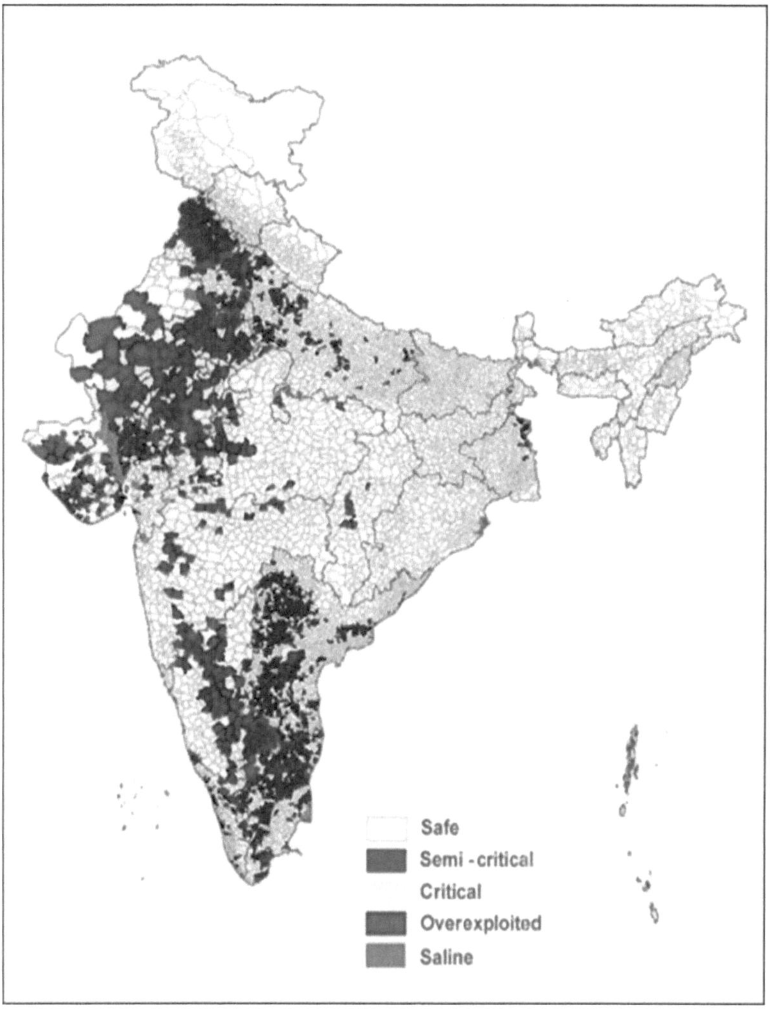

Figure 4.2 Map of groundwater exploitation, 2006.

Source: Towards Better Management of Ground Water Resources in India. B.M. Jha, Chairman, and S.K. Sinha, Chief Scientist of the Central Ground Water Board, Government of India (https://cgwb.gov.in/sites/default/files/MainLinks/GWRA-2017-National-Compilation.pdf).

the nearest stations almost daily to fill them with drinking water for their families. This prompted the central government to launch an emergency 5-year scheme to conserve and better use available groundwater in 78 particularly hard-hit districts in seven states of the country. Five of the seven states are in the three arid and semi-arid zones of the country most suited for generating wind and solar energy: Gujarat, Karnataka, Madhya

Table 4.1 State of groundwater utilization, 2006 (in billion cubic metres).

Region	Number of observations	Net recharge	Level of exploitation (%)	Overdrawn	Critical
Western	462	25.4	96	172	62
Central	985	85.5	42	31	6
Southern		70.7	61	432	126
India		399.26	58	839	226

Source: Towards Better Management of Ground Water Resources in India. B.M. Jha, Chairman, and S.K. Sinha, Chief Scientist of the Central Ground Water Board, Government of India.

Table 4.2 Land categories and area (in kilometres and percentages) in Karnataka and Tamil Nadu and rainfall (in millimetres).

State	Karnataka		Tamil Nadu	
	km^2	%	km^2	%
Total	190,500	100	130,330	100
Forest, water, and urban	45,300	23.7	43,300	33
Wasteland	32,000	16.8	35,000	27
Cropped area (net)	102,260	53.2	52,030	40
of which Irrigated agriculture	36,590	35.5	28,639	59
Rainfed agriculture	63,410	62	20,200	39
Rainfall total	1,220	100	980	100
June to December	1,036	85	820	83
January to May	184	15	160	17

Sources: State Department of Agriculture Bengaluru, Profile of Agriculture Statistics 2015–16; K. Jothi Sivagnanam: State Agriculture Profile of Tamil Nadu 2010–11. Prepared for Ministry of Agriculture, New Delhi.

Pradesh, Maharashtra, and Rajasthan. A detailed analysis of the crisis in two states in southern India, Karnataka and Tamil Nadu, Tables 4.2 and 4.3, reveals how the rising pressure on land and the exhaustion of groundwater have exacerbated the risks for uninsured dryland agriculture to create the crisis the country faces today. This interaction is evident in the other eight mainly dryland farming states as well.[9]

Table 4.2 shows that up to 85% of the rainfall in these states takes place between June and January. The remaining 5 months of the year are almost completely dry. If the north-east monsoons fail, as they did in 2018, then the dry spell starts in September and lasts for 8–9 months up to the following June. Despite this, 62% of the land under cultivation in Karnataka and 39% in Tamil Nadu are entirely dependent on the rains.[10] Farming has therefore been a high-risk operation since time immemorial, especially since one in four monsoons is weak on average, even when the monsoon does not fail.

Fragmentation of holdings

Table 4.3 shows that 18.2% of the land holdings in Karnataka are of less than 1 hectare and almost 30% are between 1 and 2 hectares. Together these account for 48% of the holdings. The fragmentation in Tamil Nadu is even more intense, but the land ownership is also more unequal: 33% of the holdings are below 1 hectare and 25% are between 1 and 2 hectares. Together these occupy 59% of the land. The average size of the marginal farm holdings in both states is below half a hectare. The average size of the next category, the 'small farm', is also truly small: under 1.5 hectares.

Table 4.4 shows how the pressure on the land has been increasing. In just 5 years, the average size of a holding decreased from 0.89 hectares in 2000–01 to 0.83 hectares in 2005–06, a decline of 6.7%. This decline has continued. Taken together, Tables 4.3 and 4.4 show that the farming community in these two states faces a bleak future. The pressure on the land and on groundwater resources cannot be allowed to keep growing indefinitely.

In 2006, 29% of the areas surveyed in the southern region of the country reported that the annual withdrawal of groundwater either exceeded or was perilously close to the annual recharge. It would be surprising indeed if this figure too has not been exceeded by 40%. In the second and third decades of the century, global warming has also begun to accelerate. The monsoons have become more erratic and the frequency and ferocity of floods and droughts in southern India have increased. So, it would be surprising indeed if the annual recharge of water to the underground aquifers has not decreased.

The condition of agriculture in the eight other states that account for nine-tenths of the dryland farming in the country is no different. Up to

Table 4.3 Land ownership pattern in Karnataka and Tamil Nadu (in hectares).

State	Karnataka				Tamil Nadu			
Size (ha)	Number of holdings	Area covered	Average size	Percentage of total land	Number of holdings	Area covered	Average size	Percentage of total land
< 1	3,849,000	1,851,000	0.48	18.2	6,230,000	2,286,000	0.37	33.5
1–2	2,138,000	3,020,000	1.41	29.8	1,230,000	1,721,000	1.39	25.2
2–4	1,267,000	3,393,000	2.68	33.4	530,000	1,468,000	2.71	21.5
4–10	511,000	2,904,000	5.69	28.7	170,000	959,000	5.64	14.1
> 10	68,000	994,000	14.71	9.8	19,000	391,000	20.58	5.7

Source: State Department of Agriculture Bengaluru, Profile of Agriculture Statistics 2015–16.

Table 4.4 Growing pressure on land in Tamil Nadu, 2000–2006.

Category	Number of operational holdings (lakh)		Area operated (lakh ha)		Average size of holdings (ha)	
	2000-01	2005-06	2000-01	2005-06	2000-01	2005-06
Marginal (< 1 ha)	58.46	62.28	21.59	22.86	0.37	0.37
Small (1–2 ha)	12.26	12.34	17.12	17.21	1.40	1.39
Semi-medium (2–4 ha)	5.71	5.42	15.51	14.68	2.72	2.71
Medium (4–7 ha)	1.93	1.70	10.94	9.58	5.67	5.64
Large (7–10 ha)	0.23	0.19	4.56	3.91	19.83	20.58
All size groups	78.59	81.93	69.72	68.24	0.89	0.83

Source: K. Jothi Sivagnanam: State Agriculture Profile of Tamil Nadu 2010–11. Prepared for Ministry of Agriculture, New Delhi.

85% of the farms in these states are below 2 hectares in size. The traditional crops they grow include cotton where there is rich black soil capable of retaining moisture for long periods, and millets (coarse grains) on the lighter soils. Aided by an army of non-governmental organizations, the governments of these states have disseminated improved seeds and a wide variety of improved agricultural practices that can increase the yield and survivability of rainfed crops. But a study of three typical villages, in Maharashtra, Karnataka, and Andhra Pradesh, in 2006, by two researchers, Jetske A. Bouma and Christopher A. Scott, from the Hyderabad-based International Water Management Institute, concluded that even where soil conservation and watershed management programmes had succeeded, farmers preferred to use their greater control over nutrient and water availability to switch to more lucrative crops such as fruit and vegetables, sugar cane, and cotton.

In general, Bouma and Scott concluded: 'the choice of dryland crops seems to be a negative choice. Farmers, who can afford to diversify, [would] rather invest in more water-intensive, high-value crops like horticulture, vegetables and oilseeds instead. Only on the more marginal, infertile and remote lands do farmers grow dryland crops. These lands tend to belong to the most marginal households, who cannot afford to invest in labour, agricultural inputs or improved seeds. Even if public investments are made in soil and water conservation, these farmers often prefer to work for wages, rather than maintain investments in their land.' Bouma and Scott warned therefore that the solution to poverty in the dryland areas lies outside agriculture: 'If under these conditions the yields of dryland crops were to be improved', they wrote, 'the most important intervention would probably be to address the root cause of poverty itself. Lack of access to productive means, poor health, poor education, high risk adversity, low wages and a weak negotiation position in village decision-making all constrain dryland crop productivity in an indirect way.'

Notes and References

1. Panagariya, Arvind (2008). *India: The Emerging Giant.* Oxford University Press, p. 153.
2. Behere, P.B. and Bhise, M.C. (2009). Farmers' suicide: across culture. *Indian Journal of Psychiatry* 51(4):242–3. https://www.ncbi.nlm.nih.gov/pmc/articles/PMC2802368/ (viewed 16/2/2022).
3. Dongre, Amol R. and Deshmukh, Pradeep R. (2012). Farmers' suicides in the Vidarbha region of Maharashtra, India: a qualitative exploration of their causes. *Journal of Injury and Violence Research* 4(1):2–6.
4. Census of Agriculture 2015–16. Table 23, p. 80. DOI: 10.5249/jivr.v4i1.68 (viewed 16/2/2022).
5. Economic Survey 2014–15. However, other estimates based on the decennial population census and NSS place the size of the migrant workforce somewhat lower, at 20–29% of the labour force of India. In 2021 the labour force was 477 million. Adding 10 million daily commuting migrant workers not included in either set of surveys, the migrant labour therefore could have accounted for 95–138 million persons. https://www.indiabudget.gov.in/budget2017-2018/es2016-17/echap12.pdf (viewed 18/2/2022).
6. Atal Bhujal Yojana: PM launches Rs 6,000 crore Groundwater Management Plan. https://www.ndtv.com/india-news/pm-launches-rs-6-000-crore-national-groundwater-management-plan-on-atal-bihari-vajpayees-birth-anniv-2154120 (viewed 18/2/2022).
7. Punjab and Haryana in the north have also overutilized their groundwater resources, but that is because these areas receive less than 800 mm of rain a year but grow two crops, mostly on groundwater irrigation. Report of the Central

Groundwater Board 2010–11. http://cgwb.gov.in/documents/papers/incidpapers/Paper%201-B.M.Jha.pdf (viewed 19/2/2022).

8. Report of the Central Groundwater Board 2015–16. https://cgwb.gov.in/Annual-Reports/Annual%20Report%202015-16.pdf, pp. 62–6.
9. This is because these states receive only the south-west monsoon.
10. This amounts to 26% of the total land area of the two states.

5

The Crisis in Irrigated Farming: End of the Green Revolution

Agriculture in the Indo-Gangetic Plain could not be more different from agriculture in the drylands of the south and west of India. While the drylands are an area of chronic scarcity, agriculture in the Indo-Gangetic Plain is beset by chronic surpluses. While the farmers of the drylands are in acute need of security for their livelihoods and protection against the inclemencies of weather, farmers in the irrigated areas have been producing chronic surpluses and have been in growing need of protection from the vagaries of the *market*.

The Indo-Gangetic Plain is India's breadbasket: 72% of the land in the five states that comprise it, Punjab, Haryana, Uttar Pradesh, Bihar, and West Bengal, is under cultivation. This means that this includes just about every square inch of land that is not already under other use. These five states provide 70% of India's rice, 90% of its wheat, and 65% of its sugar cane. Fully half of India's output of sugar cane (170.5 Mt out of 340 Mt in 2017–18) comes from Uttar Pradesh. This vast area is now caught in what has come to be known as 'the cereals trap' and does not know how to get out of it. Biomass gasification, which has so far been considered solely in the context of the role it can play in replacing fossil—specifically transport—fuels, is a technology that could provide a way out of this trap. This chapter is devoted to describing the origin of the cereals trap. The transformative role that biomass gasification could play is described in Chapters 13 and 15 and for the specific case of India in Chapters 16 and 17.

Sun, Wind, and Biomass. Prem Shankar Jha and Michael B. McElroy, Oxford University Press.
 DOI: 10.1093/9780198944768.003.0005

The Green Revolution, double cropping, and soil exhaustion

The origins lie in a relentless double cropping of irrigated land that can be traced back to the Green Revolution in Indian agriculture. The Green Revolution began in the late 1960s, coincidentally only a year after two successive failures of the south-west monsoons in 1965 and 1966 devastated the countryside and created rampant inflation in the cities. Both starvation in the villages and inflation in the cities were averted only by the gift from the US of millions of tonnes of wheat under its Public Law (PL) 480. In the winter harvest of 1968, the production of wheat jumped from 6 Mt to 11 Mt. Despite this, both the Indian and the US government took no chances and the US continued the supply of PL 480 wheat up to 1971, after which it was no longer needed. By then the US had sent 55 Mt of wheat to India, and the Green Revolution in wheat, which had begun with the crossing of Mexican with Indian varieties of cereal, had taken firm hold.

By the mid-1970s, the Green Revolution had spread also to rice. The Indian Agriculture Research Institute had developed 140 varieties of hybrid rice, tailored to the soil, water, and sunshine availability in every part of the country. In addition to their higher yield, the new varieties matured in a shorter period of time. This made it possible for farmers in the western half of the Indo-Gangetic Plain to sow and harvest a rice crop before planting their wheat crop in November, and for farmers in the eastern half to sow a wheat crop after harvesting their main, traditional, paddy (unmilled rice) crop in October.

The result was double cropping with wheat and paddy throughout the Indo-Gangetic Plain. As much as 28.6 out of the 37.5 mha of cultivated land in the five states of which it is comprised are irrigated, and all but a very small part of this land is double cropped with rice and wheat or with two crops of rice (Table 5.1).

The increase in the production of these major cereals was dramatic. The production of rice and wheat rose from 43.2 Mt and 23.2 Mt respectively in 1970–71 to 74.6 Mt and 54.5 Mt respectively in 1989–90.[1] Up to this time in the 1970s, wheat had been grown almost exclusively in Punjab, Haryana, and western Uttar Pradesh. Its spread to the east brought the country's production of wheat to a par with rice. Since the

Table 5.1 Land use in the Indo-Gangetic Plain (in million hectares).

State	Irrigated area	Dryland farm area	Total cropped area	Total area
Punjab	3.96	0.04	4.0	5.03
Haryana	3.26	0.34	3.6	4.42
Uttar Pradesh	14.49	3.11	17.6	24.1
Bihar	3.90	2.5	6.4	9.42
West Bengal	2.98	2.52	5.5	8.87
Total in Indo-Gangetic Plain	28.59	8.51	37.1	51.84

turn of the century, India's annual production of wheat has been within 10% of its output of rice.

The end of the Green Revolution

By 1990, India had achieved self-sufficiency in foodgrains. But as the output of both rice and wheat continued to grow, the Green Revolution became a mixed blessing. Half a century of relentlessly intensive cultivation has exhausted the productivity of the soil of the Indo-Gangetic Plain. The tell-tale indicator was the ever-growing use of fertilizer. While the production of foodgrains increased 1.33 times from 126 Mt in 1978 to 285 Mt in 2018–19, the annual consumption of chemical fertilizers rose more than six times, from 4.29 Mt to 27.23 Mt.[2]

Faced with a 'crisis of plenty', the central government adopted a compulsory food procurement system that had been introduced during the Second World War to feed allied troops facing the Japanese in eastern India and to stabilize the price of food in the urban areas. They converted this into a price support system in which the Food Corporation of India, a central government-owned enterprise, was obliged to buy all the foodgrains that the farmers wanted to sell at a minimum support price (MSP) and to sell it at a highly subsidized price to the urban and rural poor through controlled price 'ration' and 'free price' shops.

But while the MSP eased problems of distribution by stabilizing the price of cereals, it exacerbated the problem of overproduction, for it insulated farmers from the vagaries of the market and encouraged them to produce still more wheat and rice. This has created a growing mountain of foodgrains in government granaries, which stood at 27.7 Mt of rice and 55 Mt of wheat in April 2020, about three times the amount needed to cushion the urban and rural poor against the worst of crop failures.

The only way to keep the mountain from growing ever larger has been to export the surplus. As a result, from being a net importer of foodgrains in the early 1970s, India has become the largest exporter of rice in the world. In 2019–20 it exported 12 Mt of rice—32.5% of worldwide exports—and 6 Mt of wheat. Much of the latter was sold as cattle feed because wheat, being hygroscopic, tends to rot, becoming unfit for human consumption.

Soil salinity and waterlogging

The ecological price of this rapid spread of double cropping has been very high, for it has steadily depleted the vast underground lake on which the Indo-Gangetic Plain floats. This depletion has been most severe in Punjab, Haryana, western Uttar Pradesh, and north-eastern Rajasthan, which are the most inland portions of the plain and receive therefore the smallest amount of rain from the south-west monsoon—750 mm or less—per year. As a result, the withdrawal of groundwater in these states has reached a level that is well beyond its annual recharge.

This excessive withdrawal has resulted in the entry of increasing amounts of saline water into the aquifers. A study released by the Indian Council of Agricultural Research in October 2020 estimated that while nearly all the 147 mha of cultivated land in the country had suffered from one or another form of soil degradation, 23 mha had become saline, alkaline, or acidified and 14 mha had become waterlogged or flooded. Much of the waterlogging—around 6.74 mha—has resulted in excessive salination. This has created a hard shell on the surface that prevents the water from seeping into the ground. The study estimated that the salinized area was increasing by a tenth every year.[3] Few farmers can afford the extra cost of deep ploughing and applying gypsum to the land to revive it. So more and more of the land has gone out of cultivation.

Rice stubble burning

The double cropping of wheat and rice has dealt the land one final blow. For almost four decades since the eighties, farmers have been burning the stubble and straw left by the rice crop to clear their fields as quickly and cheaply as possible after harvesting. This is not a peculiarly Indian problem. There are photographs (Figure 5.1) of crop stubble being burned in Essex, England, in 1986, and stubble burning is a common practice in the US South-East. Rice stubble burning in Punjab is shown in Figure 5.2.

Stubble burning began in India in the 1980s and had become a near-universal practice on double and triple cropped lands by the turn of the century. As described in Chapter 6, this has not only become a serious health hazard in cities, towns, and villages that lie to the south-east of Punjab, including India's capital, New Delhi, but also dangerously accelerated the impoverishment of the soil. This is because the heat from the fires regularly kills the microbes, insects, and bacteria that fertilize

Figure 5.1 Stubble burning in north Essex.

Source: John Rostron: File: Stubble burning in North Essex—geograph.org.uk—1448777.jpg. Location: 51°46′43.82″N, 0°48′24.84″E. Created: 1 September 1986. https://creativecommons.org/licenses/by-sa/2.0/.

Figure 5.2 Rice stubble burning in Punjab.

Sources: Suhasini Krishnan, What is stubble burning? How does stubble burning cause smog? Alternatives to stubble burning, 2020, https://www.thequint.com/explainers/stubble-burning-delhi-air-pollution-smog; Neil Palmer, https://en.wikipedia.org/wiki/Stubble_burning#/media/File:NP_India_burning_48_(6315309342).jpg (viewed 4/3/2022), https://creativecommons.org/licenses/by-sa/2.0/.

the soil and also burns away nitrogen and trace minerals in the soil, such as potassium and magnesium, that nourish the crop. To compensate for this, farmers have been forced to use more and more fertilizer to obtain the same yield per hectare. Stubble burning has also been creating a hard crust, about an inch thick, that reduces the aeration of the plant roots and prevents rainwater from seeping into the soil to feed the roots of the next crop. This has accelerated waterlogging and soil salinity.

The conclusion is therefore inescapable: whether one looks at the wheat bowl of India in the north-west that is now also growing rice or the rice bowl of India in the east that is now also growing wheat, the Green Revolution has exhausted its potential. But the support structure that central and state governments have built, from large subsidies for fertilizers and pesticides to cheap, often free, electricity and assured sales at remunerative prices, has created the 'cereals trap' from which farmers have not been able to find a way out.[4]

Diversification into horticulture

Farmers began trying to break out of this trap in the early 1980s. Their way out was to branch into horticulture, dairying, and poultry farming. Horticulture, especially the growing of vegetables, is far more labour intensive than cereal farming, but also far more productive. In a typical Punjab village, for instance, in 2018–19, 1 hectare under cereals yielded 4 tonnes of paddy and 5 tonnes of wheat, but 19.93 tonnes of vegetables.[5] For understandable reasons, therefore, those with small and marginal holdings and an abundance of family labour were the first to go into horticulture. The growth in horticulture as compared with foodgrains is illustrated in Figure 5.3.

The shift into horticulture began in the 1980s, but the world the farmers entered was frighteningly different from the world they were leaving, for it was one in which near-complete market security was replaced by an equally complete market insecurity. While cereals are not perishable and can be stored easily for 6 months (wheat) or even for several years (lentils), onions and potatoes last for at most a few weeks without getting severely desiccated or rotting. More delicate vegetables such as peas and tomatoes, in addition to fruit, milk, and eggs, spoil in days.

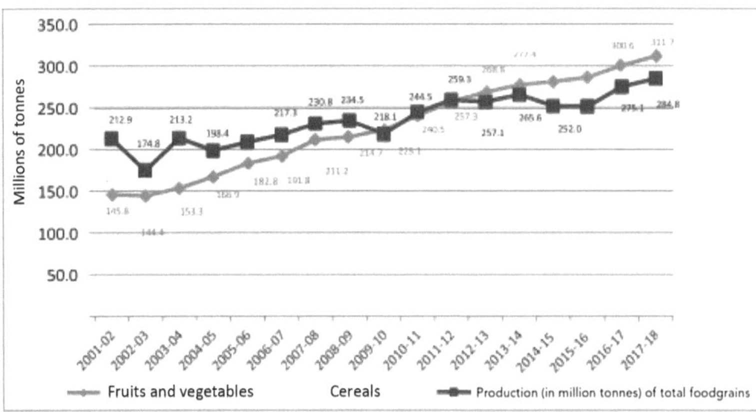

Figure 5.3 The rapid growth of horticulture vis-à-vis cereals.

Source: Government of India Controller of Publication, Horticultural Statistics at a Glance 2018: Ministry of Food and Farmers' Welfare.

It was not until 2004–05 that the central government recognized the potential of horticulture to alleviate rural poverty. They created a National Horticultural Mission (NHM) to coordinate its development with the state governments. In the decade and a half that followed, the production of fruits and vegetables rose from146 Mt to 324 Mt (2019–20). The NHM made this possible by helping create an urban infrastructure for storing, marketing, and, above all, exporting fruit and vegetables from India. In March 2019, there were an estimated 7,645 large cold storage facilities in India with a refrigerated space of 150 million cubic metres, capable of storing 37–39 Mt of perishable produce.

By 23 September 2020, this number had risen to 8,186. But nearly all of these were in the cities and larger market towns and concentrated on less-perishable vegetables, such as onions, potatoes, and processed dairy products, destined for export. As much as 68% of the cold storage units stored only potatoes. Most of the remaining 32% are dedicated to meat and dairy products and onions, all of which have become important exports for India.[6]

As Figures 5.4 and 5.5 show, peak prices for all three vegetables in the wholesale market are on average double the corresponding floor prices. This is the case for the least-perishable vegetables. The gap is even greater for peas, cabbage, cauliflowers, and other more delicate vegetables. But the actual growers of the vegetables get at most half of the wholesale price. While there is no all-India study of what these 'farmgate' prices are, a study of the all-India average wholesale prices of onions in 2013 and 2014 by the Indian Council for Agricultural Research (ICAR) showed that in those years the price was Rs 12.65 and Rs 11.03 per kg respectively.[7] Farmers seldom, if ever, got more than half of this price from the wholesalers. So, it would be reasonable to assume that they received between Rs 5.52 and 6.32 per kg respectively. But, as is shown in Figure 5.5, the all-India average *retail* price of onions peaked at Rs 58 per kg in November 2013 and at Rs 55 per kg in October 2015, before being forced down by hastily imposed price controls. So, the actual growers got as little as a tenth of the average market price for their onions when they took their crop to the market. The gap is even greater in tomatoes and still greater in vegetables such as peas, cabbage, and cauliflower with a shorter shelf life. It is greatest of all in floriculture.

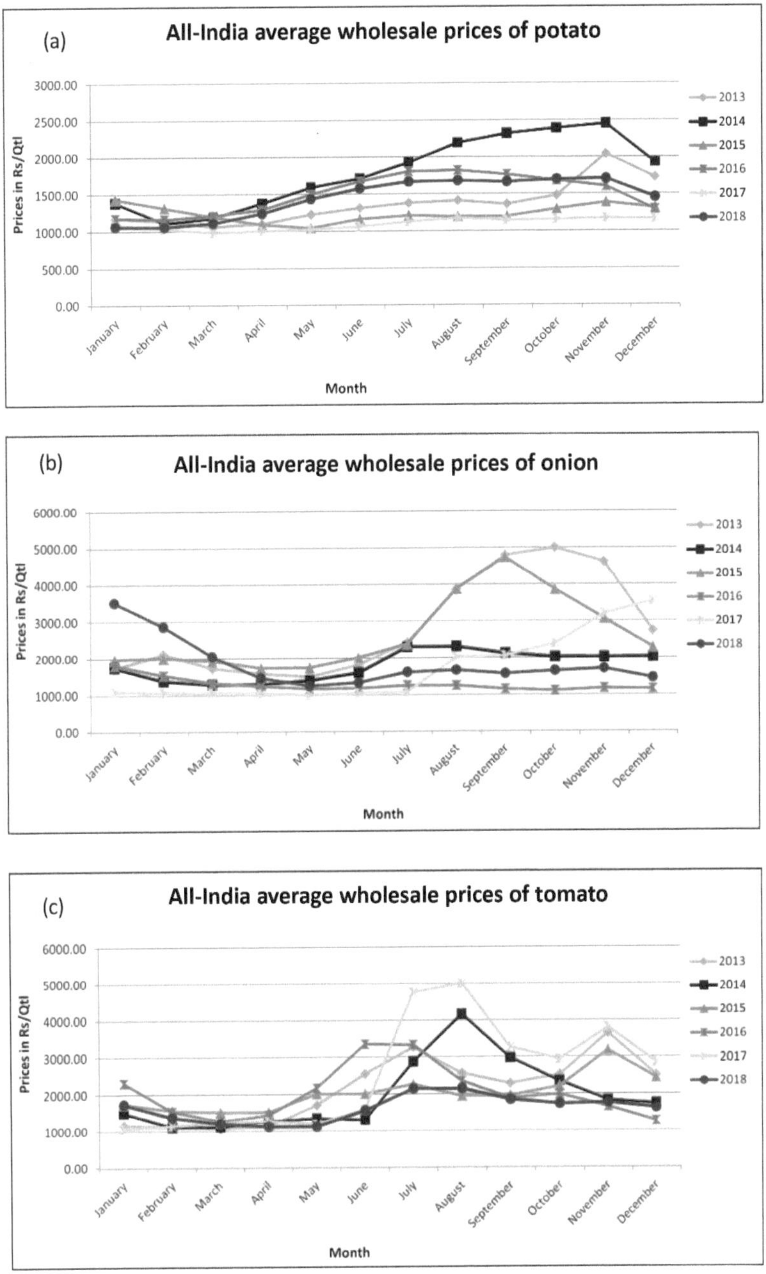

Figure 5.4 All-India average wholesale prices of potatoes, onions, and tomatoes, 2013–2018 [1 quintal (Qtl) = 100 kg].

Source: Government of India Controller of Publication, Horticultural Statistics at a Glance 2018: Ministry of Food and Farmers' Welfare, pp. 331–6.

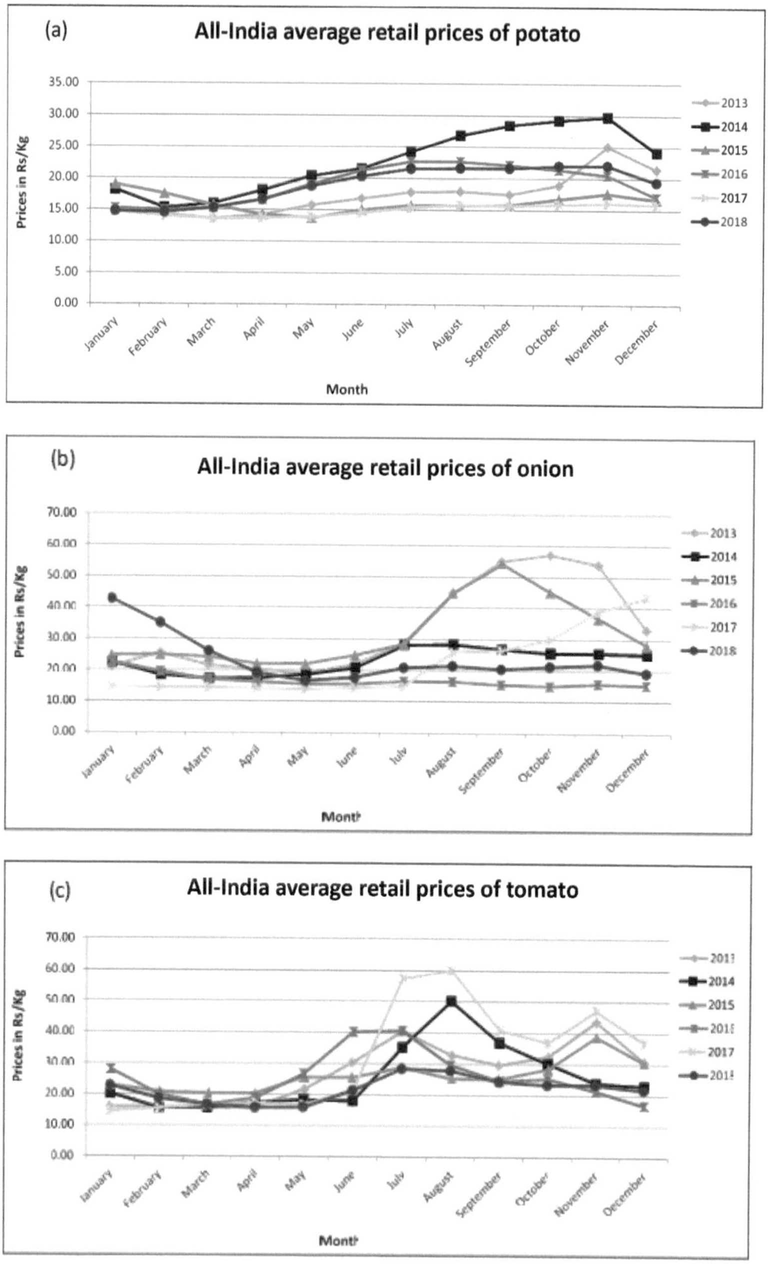

Figure 5.5 All-India average retail prices of potatoes, onions, and tomatoes, 2013–2018.

Source: Government of India Controller of Publication, Horticultural Statistics at a Glance 2018: Ministry of Food and Farmers' Welfare, pp. 331–6.

The retail markup over 'farmgate' prices of onions, potatoes, and tomatoes would have been even greater had these prices not been moderated since the early 2000's by the pre- emption of a part of the crop by exporters at the very beginning of the marketing season which gave farmers a second avenue of sale. This has, to some extent, raised and stabilized the minimum price that farmers are able to secure but increased the vulnerability of domestic consumers who now have to bear the entire brunt of the shortfall in production caused by unseasonal rain or pest attacks later in the season. Whenever this has happened, as it did most dramatically in the Nasik district of Maharashtra, the main onion growing region of the country, in December 2014 and again in November 2019,[89] the government has been forced to react by imposing direct controls on retail prices to limit the misery caused to the urban poor. This intervention is visible most clearly in the sharp break and subsequent flat-line drop of onion prices in 2015.

The perishability of their produce has thus placed fruit and especially vegetable farmers at the mercy of the traders. For once they have harvested the crop, they must sell it at any price or watch it perish. This absence of bargaining power has ensured that vegetable, dairy, and poultry farmers receive between a quarter and a tenth of the market price for their produce, depending on its perishability and the conditions prevailing in the market. In bumper crop years, this barely covers the cost of transport.[10]

Despite these uncertainties, the shift out of cereals into horticulture has continued. The area under horticulture grew from 12.8 mha in 1991–92 to 25.4 mha in 2017–18. The reason for this was a rapid and unexpected growth of exports. As the exporters offer contracted prices to ensure, and often pre-empt, supply, this provided a degree of stability to farmers' earnings. From $2 billion (Rs 80 billion) in 2000–01, exports quadrupled to $8.5 billion (Rs 637 billion) in 2018–19.

To manage the growing volume of horticultural produce, cold storage facilities have sprung up all over India. As mentioned earlier, in March 2019 there were an estimated 7,645 large cold storage plants with a refrigerated space of 150 million cubic metres, capable of storing 37–39 Mt of perishable produce. But these catered mainly to the export trade, and, even in this, concentrated on the less-perishable vegetables, such as onions, potatoes, and processed dairy products. So, 68% of the cold storage facilities stored only potatoes.[11] In Punjab, for example, which is the sixth largest producer of potatoes in the country and where most of the crop consists of highly valued seed potatoes that its farmers sell to potato farmers in southern India, 362 of 378 of cold storage facilities in 2018 stored only potatoes. All of these were in urban areas. Another 7 stored only milk products. That left only 9 multi-purpose cold storage plants. *Not one of these was located in a village.*

The pattern is the same in the rest of India. Most of the remaining 32% of the all-India cold storage space is dedicated to storing dairy products and onions, both of which have become important exports for India. There are still very few cold storage facilities in the villages that are equipped to handle perishable vegetables such as peas and tomatoes, whose shelf life is a good deal shorter. As a result, farmers who grow these prized vegetables have no option but to hire trucks or pick-ups and deliver their produce directly to the city wholesale markets. Once they are there, the perishability of their produce leaves them with no bargaining power and they are forced to take whatever price the wholesaler offers. Ever-increasing supplies of fruit and vegetables to the market have also progressively increased the risk that middlemen run of not being able to sell the produce before it rots. The facilities have reduced this risk, but the more the farmers produce, the greater becomes the risk the middlemen run of not being able to sell all of what they bought from the farmers. According to several estimates, between them the farmers and the middlemen are unable to sell up to 30% of vegetable and fruit crops, valued at Rs 440 billion (almost $6 billion), either because they are damaged during transport or because they begin to rot before they can be sold.[12] For the middlemen, the risk rises with the size of the crop, so they try to minimize it by lowering their offer price to the farmers. For the farmers therefore this is a 'heads you win, tails I lose situation'.

Notes and References

1. The 1970–71 data are from a US Central Intelligence Agency document. https://www.cia.gov/library/readingroom/docs/CIA-RDP85T00875R001700020049-5.pdf (viewed 2/3/2022).

2. Fertiliser Association of India: Fertiliser Statistics 2019. https://www.faidelhi.org/general/con-npk.pdf (viewed 2/3/2022).
3. Kumar, Pardeep and Sharma, Pradeep K. (2020). Soil salinity and food security in India. *Frontiers in Sustainable Food Systems* 4.
4. In November 2020, with bulging granaries and no way to stop them from growing even larger, the Indian government fell back on the only option that remained: it abolished the MSP system which had been administered by the state governments and threw the market open to grain traders from all over the country and all over the world. For the wheat and rice growers of the entire country this felt like they had been pushed off a high cliff into the sea. A violent farmers' resistance developed in the country with, as of December 2024, no end in sight. https://thewire.in/agriculture/farm-laws-protest-narendra-modi-bjp-government-familiar-pattern (viewed 5/3/2022).
5. The nationwide average between 2018 and 2021 was just over 18 tonnes per hectare.
6. Virendra Singh Rawat, *Business Standard*, 23 March 2019. https://www.business-standard.com/article/economy-policy/cold-storage-capacity-expands-rs-21-000-crore-investment-lined-up-by-2023-119032200623_1.html (viewed 8/3/2022).
7. ICAR: State-wise analysis of production, storage, market arrival and price trends for onion and potato. https://agricoop.nic.in/sites/default/files/ICAR_5.pdf (viewed 8/3/2022).
8. Unseasonable rain, hailstorm wreak havoc on cash crops in Nashik, *The Economic Times*, 13 December 2014. https://m.economictimes.com/news/economy/agriculture/unseasonal-rain-hailstorm-wreak-havoc-on-cash-crops-in-nashik/articleshow/45504313.cms (viewed 9/3/2022).
9. Rain damages kharif onions on 17,500 hectare, *Nashik News, The Times of India*, 6 November 2019. https://timesofindia.indiatimes.com/city/nashik/rain-damages-kharif-onions-on-17500-ha/articleshow/71929562.cms (viewed 10/3/2022).
10. Devidas Parmane, a farmer from Vadgaon Rasai village near Pune, sold 952 kg of onion in Pune Agricultural Price Market Committee on 10 May 2021 for Rs 1.60 per kg. After deducting the market charges and transportation cost, he received a net payment of Rs 1 per kg. https://economictimes.indiatimes.com/news/economy/agriculture/re-1-thats-what-farmers-get-for-a-kg-of-onion/articleshow/52408624.cms?utm_source=contentofinterest=&utm_medium=text=&utm_campaign=cppst= (viewed 11/3/2022).
11. Virendra Singh Rawat, *Business Standard*, 23 March 2019. https://www.business-standard.com/article/economy-policy/cold-storage-capacity-expands-rs-21-000-crore-investment-lined-up-by-2023-119032200623_1.html (viewed 11/3/2022).
12. There are different definitions of crop wastage and different crops in the baskets referred to; 18% is a reliable minimum estimate. It is obtained from the Centre for

Public Policy Research paper 'Government's Role in India's Ailing Cold Storage Sector', December 2016, by Madhu Sivaraman. https://www.cppr.in/wp-content/uploads/2016/12/Government%E2%80%99s-Role-in-India%E2%80%99s-Ailing-Cold-Storage-Sector.pdf (viewed 11/3/2022). However, several other estimates put the overall wastage at 25–30%. See Dilip Kumar Jha, *Business Standard*, 19 April 2020. https://www.business-standard.com/article/economy-policy/farmers-let-rabi-30-veggies-rot-in-fields-start-ploughing-for-kharif-120041800681_1.html; and Nath, A., et al. (2018). Postharvest management of horticultural crops for doubling farmer's income. *Journal of Pharmacognosy and Phytochemistry* 2018:2682–90. https://www.phytojournal.com/archives/2018/vol7issue1S/PartAN/SP-7-1-708.pdf (viewed 15/3/2022).

6
Coping with Air Pollution

As will be outlined in this chapter, the problems associated with the deterioration of air quality in India are not uniformly distributed across the subcontinent. To guide in the interpretation of geographic features of the data, we begin by displaying a map of the names and spatial distributions of the states that compose the national territory.

The Indian government identified six major pollutants of particular concern in terms of their impact on public health. Most significant, as we shall see, are small particles that can be respired entering the lungs, triggering a variety of negative health effects. These particles are identified in two size ranges: PM10 with diameters less than 10 micrometres (10 μm) and PM2.5 with diameters less than 2.5 μm. The smaller particles are particularly hazardous in that they are capable of entering the bloodstream, resulting in a variety of cardiovascular complications. The other pollutants highlighted by the Indian government include ozone (O_3), nitrogen dioxide (NO_2), sulphur dioxide (SO_2), and carbon monoxide (CO).

The significance of a particular pollutant is identified by a number referred to as the air quality index (AQI), with a composite value intended to define the overall state of the air quality, typically an average over some time interval, 24 hours for example. The definitions of the various AQI ranges and their implications for public health are summarized in Table 6.1.

India suffers from all of the diverse problems of air pollution that have plagued countries in the West and more recently China. Many of these problems are directly related to the use of fossil fuel: in the domestic and commercial sectors, in transport, in industry, and in the generation of electric power. A successful transition to a low carbon future could be expected therefore to seriously cut emissions from all of these sources. But India has an additional problem that arises from the seasonal burning

Sun, Wind, and Biomass. Prem Shankar Jha and Michael B. McElroy, Oxford University Press.
 DOI: 10.1093/9780198944768.003.0006

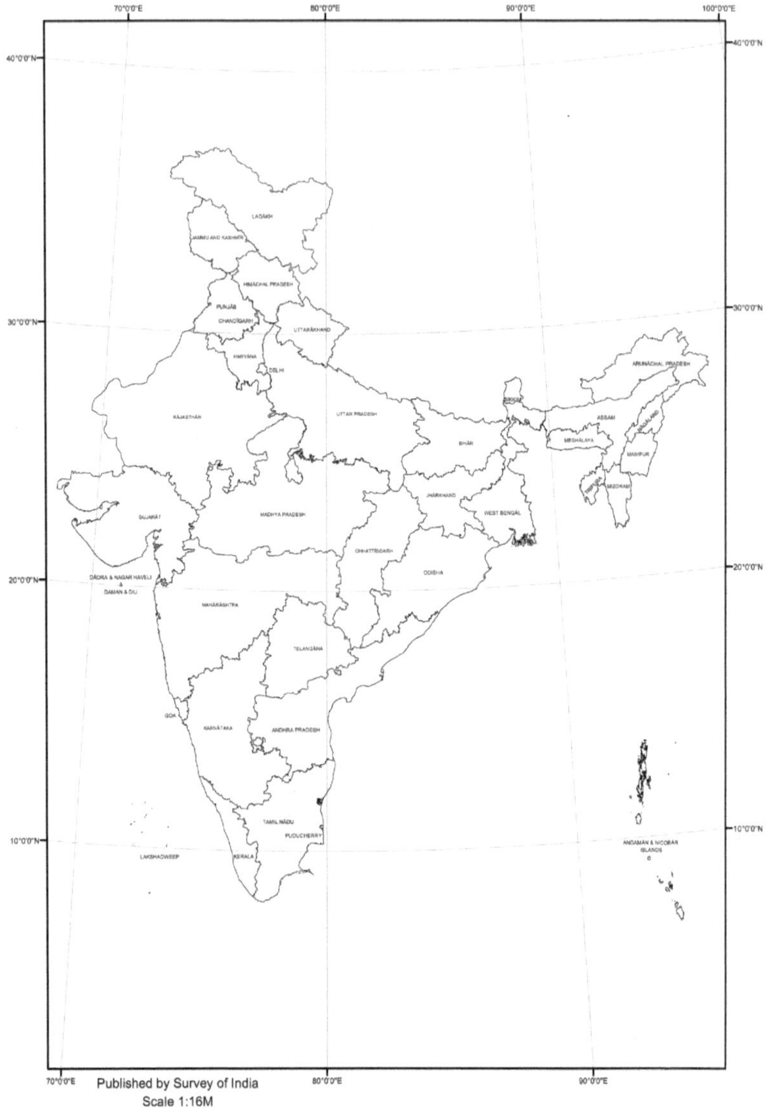

Figure 6.1 Map of India.
Source: https://surveyofindia.gov.in/webroot/UserFiles/files/1_16-state%20boundary-1.pdf.

of agricultural stubble in the fields. The sensitivity of this issue was highlighted by the reaction of farmers in Punjab and Haryana, the two main foodgrain surplus states in the country, to a decision by the state-owned Food Corporation of India, the entity responsible for running the

Table 6.1 AQI categories and their implications.

AQI category range	Implication
0–50	Good
51–100	Satisfactory; minor problems for sensitive individuals
101–200	Moderate; breathing problems for people with respiratory complications
201–300	Poor; breathing problems for people with prolonged exposure
301–400	Severe; can lead to respiratory issues
> 400	Hazardous; respiratory problems for even healthy individuals

national subsidized foodgrains sales system, to change the date for paddy rice procurement in 2021 from 1 October to 11 October.

As reason for the delay the central government cited the presence of excessive humidity in the rice grain, caused by heavy unseasonal rainfall in September. These rains had increased the moisture content to 18.2–18.7%, as compared to the prescribed maximum of 17% the government planned to buy. Another 10 days of drying out, the government's experts had concluded, would eliminate the excess. But the farmers were convinced that the central government's real purpose was to force them to sell their paddy to private traders, and to thus bring in by the back door the privatization of the grain trade upon which the government had set its heart 2 years earlier.

Why did a mere 11-day delay in the start of procurement matter so much? The answer is that farmers who grow both rice and wheat have only a 20- to 25-day window in which to clear their fields of the rice stubble left by the paddy harvest before planting the wheat. The delay would therefore allow them only 9–14 days in which to clear their fields. This would leave many with no option but to sell to private traders immediately at a discount or to burn the stubble to clear the fields for planting the wheat crop immediately after bringing in the rice harvest. But burning paddy straw had been banned since 2015 and had been attracting increasingly severe fines. So the farmers found themselves trapped between a rock and a hard place. Setting fire to the stubble without even trying to find some other use for it was therefore the only available option.

In 2020, Punjab and Haryana had planted 3.1 mha with rice varieties that needed immediate harvesting.[1] These fields produced around 18 Mt of paddy straw. Punjab and most of Haryana lie due north-west of Delhi, the direction from which the prevailing winds blow from October to May. A belated recognition of what the smoke from the stubble burning would do to the already unhealthy air of Delhi may have been responsible for the government to hastily withdraw its earlier order and allow procurement to start from 3 October.

The realization that paddy straw burning had become a serious health hazard has surfaced only recently. The crystalline blue skies of Delhi's winter had begun to shade into grey as far back as the mid-1980s. By the early 1990s, 2- and 3-day episodes of light smog with cloudless but grey skies and a wan sun that provided light but no heat had become increasingly common.

The worsening was not sudden. A 3-year study by the World Bank Development Research Group conducted from 1991 to 1994 had found that the average total suspended particulate (TSP) level in Delhi's air had risen to approximately five times the World Health Organization (WHO)'s estimate of the safe maximum level. What was worse, it had exceeded the safe daily maximum for 97% of the time on all of the days when readings were taken.

A follow-up report compiled by the Ministry of Environment in 1997 had put the blame for the deterioration on unregulated construction, antiquated small-scale industries, coal-fired power plants, and an increase in vehicular traffic. The report estimated that Delhi emitted about 3,000 tonnes of air pollutants every day. Of this, motorized transport was judged responsible for 67% of the total, coal-based thermal power plants for 12%, and industry for 12%. More alarmingly, the report noted that emissions of carbon monoxide from vehicles had almost doubled (increased by 92%) between 1989 and 1996.[2]

By 2015, some among Delhi's elite had begun to wear the white surgical masks that were then common in Beijing. The US Embassy was reported to have purchased 1,800 air purifiers for its staff members' homes. Several other embassies had followed suit. Indian companies had begun also to order filtration systems for their office buildings.[3]

But officials held fast to the belief that the worsening air pollution problem was mainly an urban issue, the result of unplanned urbanization and 'dirty' economic development, and that its onset was not significantly different from what had been experienced by the industrialized countries of Europe and the US at an earlier stage of their development.

This comforting belief was reinforced inadvertently by another benchmark study, conducted by two professors of civil engineering from the Indian Institute of Technology in Kanpur. The authors of this study used data collected in the winter of 2013 and the summer of 2014.[4] Using information from six sites spread evenly across residential, industrial, commercial, and mixed neighbourhoods in Delhi, they concluded that the load of particulate matter, both PM10 and PM2.5, had remained four to seven times higher than the acceptable maximum during summer and winter respectively. Averaging over the two seasons, they concluded that vehicular emissions and road dust accounted for 56–60% of the problem, cooking fuels for about 22%, and power plants and industry for the remaining 20–22%.[5]

The report pointed out, however, that there was a significant difference in the composition of the particulate matter as observed between summer and winter. Silicon, aluminium, iron, and calcium, the principal components of soil, road dust, and airborne fly ash, accounted for about 40% of PM10 in summer. In winter their share dropped to a mere 13%.

The contrast between summer and winter was even more marked in the composition of PM2.5. The share of soil, road dust, and fly ash dropped from 20% in summer to a mere 3.5% in winter, while the contribution of carbon produced from biomass burning and the contribution of particles containing nitrate, sulphate, and ammonia rose to 23 and 28% respectively, accounting for 51% of particulate matter in the atmosphere.

As for the cause of this vast discrepancy, the authors of the report highlighted the presence of large amounts of potassium in the particulate matter and stated that this came from the burning of biomass. They pointed out also that the potassium level was unevenly distributed over the winter months, but that 'The highest potassium levels (~15 $\mu g/m^3$) were seen in the beginning of November and early winter perhaps due to crop residue burning in Punjab and Haryana'.

This was the first report that put much of the blame for the increasingly severe air pollution in winter on a cause that lay outside the city. But the timing of its release was unfortunate, for the authors presented it to the government in January 2016. By then, the winds that heralded the waning of winter had started to blow and pollution levels had started to fall. So its reference to an external cause for Delhi's air pollution in winter was lost in the mass of data and acronyms that the authors used in their report, and the report failed to get the attention it deserved. The comforting belief that India was travelling down a road the West had already travelled continued therefore to prevail. The headline on the Kanpur study in the Delhi edition of *The Economic Times*, India's most influential Business daily, said it all: 'IIT-Kanpur study says trucks and road dust are bigger pollutants than cars in Delhi'.[6]

The complacency did not last long, because Delhi received its second shock later in the year, on 31 October 2016, when its residents woke up to find themselves enveloped in a toxic smog that lasted for 6 days (Figure 6.2). In that week, the particulate matter concentration in the atmosphere reached 999 and went off the chart in several parts of the city. This was *15 times* the maximum safe level prescribed by the WHO. The Delhi state government closed all schools, stopped all construction for a week, and deployed bulldozers to shovel mud onto two hills of burning

Figure 6.2 What Delhi looked like on 31 October 2016 at 14.37 hours.
Source: Published by *The Guardian*, 31 October 2016.

garbage that lay just outside the city limits to the north and east of the capital.[7]

Since then, the November smog has become an annual, dreaded phenomenon. It arrived in almost the same weeks in 2017, 2018, and 2019. In 2017, the PM2.5 level for the city touched 640, 10.7 times the prescribed maximum.[8] In 2018 it exceeded the WHO-defined maximum level in some parts of the city by factors as large as 20. This again led to the closing of schools, the interruption of construction, and a ban on entry of diesel-fuelled trucks into the city, and these constraints were imposed for an even longer period.[9] Regular satellite surveillance by the Indian Agricultural Research Institute detected 65,408 burning events in Punjab, Haryana, and Uttar Pradesh between 30 September and 15 November. The straw in over half of these, 39,618, was set ablaze in just 2 weeks, between 1 and 15 November. By 15 November, more than 80% of the burning had happened in Punjab. Punjab, Haryana, and Uttar Pradesh are not the only states, however, in which this burning occurs. Satellite imagery identified more than 2,000 sites of rice straw burning in 2018.[10] The map in Figure 6.3 shows that while crop residue burning took place in October all over the country, by far the largest concentration was in the north and particularly in the north-west, the direction from which winds blow mainly throughout the winter.[11]

In 2019, the average PM2.5 again reached 613, about the same level as in 2017, but it was above 800 in several parts of the city for almost a week from the beginning of November. The now routine closing of schools, construction sites, coal-fired power plants, and diesel truck transport followed.[12]

In 2020, IQAir, the US-based air pollution monitoring agency, listed 35 Indian cities among the 50 year-round most polluted cities in the world.[13] Of these, 18 are in the rice–wheat double cropping region of northern India and Pakistan's adjoining Punjab province. The data for the top 10 cities are summarized in Figure 6.4.

In the media, an awareness that air pollution in India's cities was not a carbon copy of the experience of the West dawned only after the occurrence of a second exceptionally severe smog in 2017, labelled now the 'Great Smog'. Among its eight principal causes, *The Times of India* assigned top place to the burning of rice stubble in adjoining Punjab and

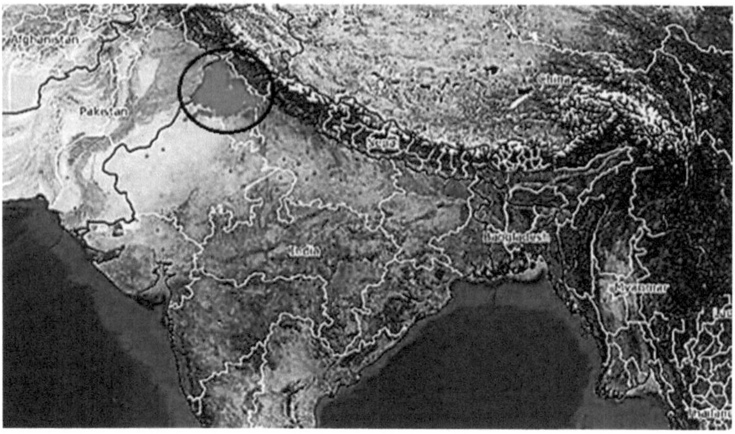

Figure 6.3 Image generated on 26 October 2018 at noon using NASA's Earth Observing System Data and Information System, showing around 2,000 spots where stubble burning is likely taking place.

Note: Not to scale; for representational purposes only.

Source: reproduced with permission from Dr Ravindra Khaiwal; https://shodhganga.inflibnet.ac.in/.

Haryana, compounded by the coincidence with Diwali, India's annual festival of lights, which is celebrated by setting off billions of firecrackers.[14]

The date of the Diwali festival is determined by the lunar calendar and falls every year between the end of October and the middle of November. These are precisely the weeks when rice stubble burning reaches its peak. The huge addition to pollutants that occurred in the 2–3 weeks that spanned the two events is locked then in place above the city by the post monsoon absence of breezes and the presence of thermal inversions that set in every winter as the land becomes colder than the air above it. Thus, the smog created by the stubble burning and the Diwali fireworks-fest lasts until the northward return of the sun begins to warm the land and freshen the wind once more.

In sum, air pollution has different origins and poses qualitatively different problems for India and South Asia than it did for the countries of Europe and the US at a similar stage of their industrialization. It is highly seasonal, concentrated mostly in winter, and increases by far more than can be explained by the enhanced burning of coal and biomass in the cities to provide winter warmth. Furthermore, the pollutants are created

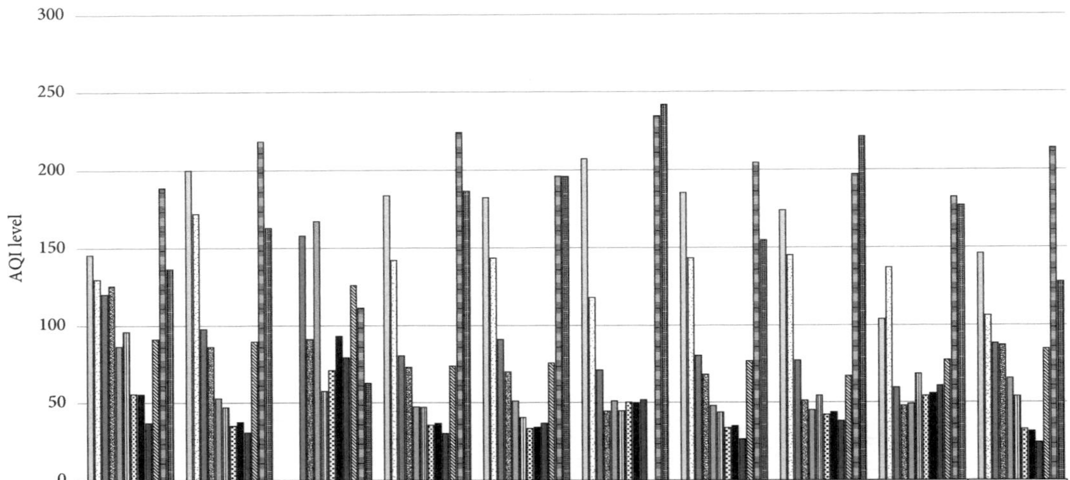

	Bhiwadi, India	Ghaziabad, India	Hotan, China	Delhi, India	Jaunpur, India	Faisalabad, Pakistan	Noida, India	Bahawalpur, Pakistan	Peshwar, Pakistan	Bagpat, India
Months										
Jan	145.8	199.9		183.7	182.2	207.1	185.3	173.9	103.9	146.1
Feb	129.8	172.2		142.2	143.5	118	143.4	145.1	137.2	106.3
Mar	120.2	97.8	158	80.5	91	71.2	80.5	77.3	59.6	88.3
Apr	125.7	86.3	91.1	72.9	70	44.6	68.2	51.4	48	87.1
May	86.5	52.9	167.4	47.4	51.1	51.2	48	45.2	49.2	65.7
Jun	95.9	47.2	57.4	47.1	40.7	44.7	43.8	54.4	68.5	53.9
Jul	55.6	35.3	70.9	35.6	33.5	50.4	33.9	42.2	54.3	32.8
Aug	55.4	37.6	93.2	36.9	34.2	50	35	43.7	55.8	31.5
Sep	37.1	30.8	79.3	30.2	36.8	51.9	26.5	38.2	60.8	24.3
Oct	91.1	89.7	126.1	73.7	75.7		76.9	67.2	77.5	84.7
Nov	188.6	218.3	111.5	224.1	196	234.5	204.4	197	182.3	213.6
Dec	136.6	163	62.6	186.4	195.7	241.7	154.8	221.2	176.9	128.1

Figure 6.4 The world's 10 most polluted cities.

Note: Vertical bars represent data for the months January to December for each region. Any blank spaces indicate that data for the month are missing, as also indicated in the table below.

Source: Created by the authors based on https://www.iqair.com/world-most-polluted-cities.

not only by industry and by transport in the cities but by agriculture, up to several hundred kilometres away.

The consequent lack of control over its causes goes a long way towards explaining the intractability of the problem. It explains also why the three countries with the worst air pollution in the world are all located in South Asia: Bangladesh with year-round AQI of 97, Pakistan with 74, and India with 72, in 2019.[15] The common feature of all three economies is that they are largely rural, with dynamic agriculture in which multiple cropping, with rice and wheat or in some places with two or even three crops of rice, is the norm. As a consequence, these countries burn a significant portion of their crop residues.

EPCA: CSE study of causes

The singularity of this pattern was highlighted in a study by the Environment Pollution Control Authority (EPCA) of Delhi and the Centre for Science and the Environment (CSE). The study correlated the daily aerosol concentration in the air (PM2.5) with wind speeds and other external stimuli, from 1 October 2017 to 25 February 2018. It showed that the PM2.5 count started below 100 on 1 October, with a sudden jump to 220 around 6 October, declining gradually again until a second, much bigger jump to 375-80 occurred on 18–20 October during the Diwali festival, falling again, albeit far less than before, to stay in the 180–200 range up to 8 November, when it shot up overnight to 670 and stayed in the 220–470 range for the next week. From then until almost the end of January, the PM2.5 concentration averaged around 200, fluctuating only with the speed of the wind. It began to decline once more only towards the end of the month as the winds freshened, and ended in February at a moderately polluted level of 100 or thereabouts.

The inverse correlation of the pollution level with wind speed displayed a near perfect relationship, except at the beginning of October when the jump in pollution coincided with a rise in wind speed. This was almost certainly the time when the dying south-east monsoon winds gave way to winds from the north-west. It was this shift, even more than the Diwali festival, that brought very large amounts of smoke and primary pollutants to Delhi from the burning of paddy stubble in Punjab and

Haryana—for the paddy harvest had begun, as it does every year, in early October. The study showed a precise, inverse correlation throughout this period between the AQI and the average daily wind speed. It concluded thus, without ambiguity, that the solution to the air pollution crisis lay both in banning firecrackers during Diwali and in ending the burning of paddy stubble.[16]

Insights from the COVID-19 pandemic

In 2020, the nationwide lockdown from 22 March dramatically improved the quality of the air throughout the country. Satellite-based aerosol density maps prepared from readings taken between 29 March and 5 April showed that pollution had declined in all but a few pockets of the country. What is most notable about the decline is that it was most marked immediately south of the Himalayas in the eastern half of the Indo-Gangetic Plain. This is the least industrialized and least urbanized but most intensive agricultural region of the valley. The sharpness of the immediate decline in this area is therefore added evidence that the winter smog has its origin to a significant extent in agriculture, and that it creates air pollution problems not only for urban but also for rural regions. The four figures from 2016 to 2019 in Figure 6.5 show the normal pattern of smog generation; the two figures from 2020 are a direct product of the COVID-19 lockdown, and they show the impact of the restrictions on human movements and paddy cultivation and the subsequent burning of straw that resulted from it.

Clinching evidence of this was the return of the smog, as if nothing had happened, on 4 November 2020. Five days after its onset, the AQI was still reading 'severe' at 474. What is more, it was above 400 not merely in Delhi but in all of the satellite cities that together with Delhi form the National Capital Region.[17] The AQI peak of 474 was still considerably below the level it attained in the previous year. But the difference was not caused by the lockdown and subsequent restrictions on travel and work. The AQI was lower because Diwali came after and not before the arrival of the smog. On 9 November 2020, the Festival of Lights (and firecrackers) was still 6 days away. By contrast, in 2019, Diwali was celebrated on 27 October, a week before the smog set in. In 2017 it occurred

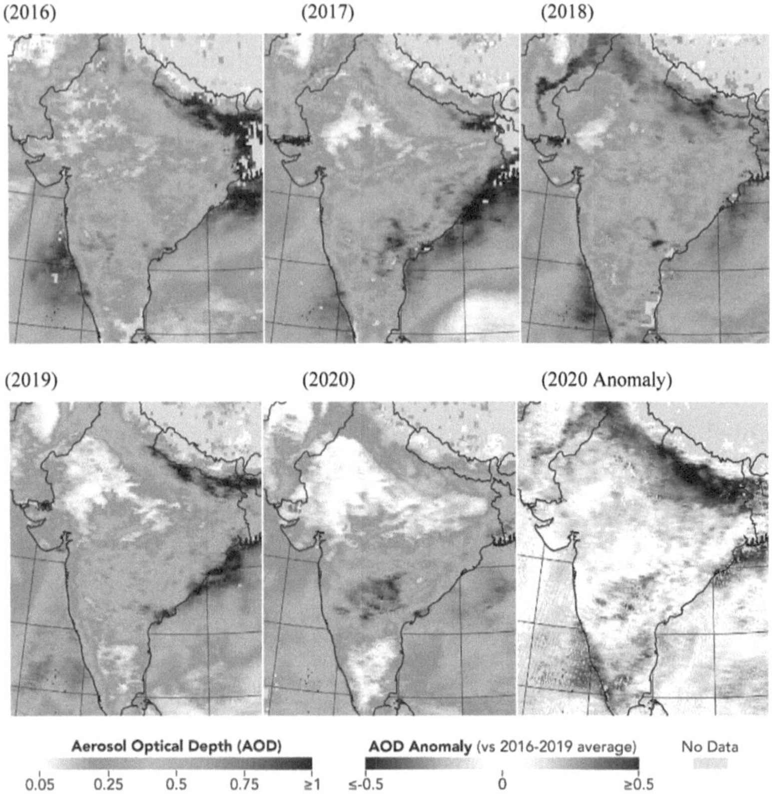

Figure 6.5 Aerosol Optical Depth, 2016 to 2020.

Note: Not to scale; for representational purposes only.

Source: As depicted by NASA. https://eoimages.gsfc.nasa.gov/images/imagerecords/146000/146596/india_tmo_2016-2020_lrg.png.

on 15 October, more than a week earlier, and in 2018 it developed on 7 November, more or less coinciding with the arrival of the smog. Thus, the optical density of the aerosol cover over Delhi in November 2020 was not significantly different from what it had been, *sans* Diwali, in the previous 3 years. Not by coincidence, 2020 was a bumper year for the paddy harvest in north India.

The COVID-19 pandemic has provided thus additional evidence that while dust from construction sites and unpaved roads, vehicular exhausts, and emissions from power stations and industry are responsible for a fairly high year-round level of pollution, the lethal smog episodes of winter in north India are the product of crop residue burning, compounded by thermal inversions and the absence of wind.

The wrong target

Recognizing that the origin of India's most severe air pollution problem arises to a significant extent from its agricultural as distinct from its industrial development opens the way for a reappraisal of what past governments have done to limit the increase of pollutants in the air. A cursory assessment shows that their record has not really been that poor. In the transport sector, leaded petrol was banned and the permissible compression ratios of automobile engines were lowered in the late 1980s; two-stroke engines that burned a mixture of petrol and oil were banned in the mid-90s and have completely disappeared from Indian roads. Blending of ethanol to replace chemical anti-knock agents and introduction of 'green' gasoline began in 2003. Standards for engine compliance with environmental norms have been raised regularly and are only slightly behind European standards. To reduce carbon emissions from passenger cars, the Modi government aims to increase blending of ethanol with gasoline to a level of 10% by 2022 from the current 8.5%, rising to 20% by 2025.

In power generation, the share of natural gas and renewables has risen sharply. By January 2021, Delhi was drawing only 8% of its power from 'unclean' sources, i.e. coal. This was a steep drop from one-third just 2 years earlier. This was obtained by shutting down two coal-based plants that had been set up decades earlier within the Delhi metropolitan area and switching two other plants in central India, which supplied most of the additional power, from coal to natural gas. About 14% of the power consumed in 2021 came from renewable energy sources.[18] City authorities have also laid down and are enforcing norms for construction that require segregation of construction sites with nylon sheeting and covering or spraying of rubble, including in excavated areas, to minimize dust.

These measures have been welcome, but their impact has gone unnoticed because of the rapid growth of the city and the annual winter invasion of smog from the countryside. As a result, what has also gone unnoticed is that pollution from vehicular exhausts, which is a serious

Table 6.2 Contributions to air pollution, 4–5 October 2021.

City	$PM_{2.5}$ (48 h max/ min)	O_3 (ppb)	NO_2 (ppb)	SO_2 (ppb)	CO (ppb)
Chandigarh	***260/72***	15	7	0	0
Delhi	***180/67***	5	16	4	8
Kanpur	*135/44*	2	0	2	2
Lucknow	*157/38*	12	7	3	6
Patna	*153/59*	1	23	2	5
Kolkata	*160/67*	11	15	7	4
Jaipur	*156/36*	7	7	9	5
Bhopal	*159/46*	9	7	32	11
Ahmedabad	*155/25*	29	19	14	0
Hyderabad	*163/74*	7	21	2	8
Bengaluru	**85/24**	–	9	4	28
Mumbai	113/63	10	18	4	18
Chennai	**96/47**	7	6	3	2

Note: The above chart has been constructed from AQI estimates for Indian cities given in World's Air Pollution: Real-time Air Quality Index (https://waqi.info/India) and follows the system used internationally. Bold formatting corresponds to good, italics to moderate, and bold italics to hazardous for sensitive groups.

concern in industrialized countries, is not the only contributor to air pollution in the cities of India. Table 6.2 shows that levels of ozone, nitrogen oxides, nitrous oxide, sulphur dioxide, and carbon monoxide between 4 and 5 October 2021 were all within safe limits not only in Delhi but also in the majority of the major cities of India. Of the 13 cities for which data are presented here, the first six, from Chandigarh to Kolkata, are located in the Indo-Gangetic Plain. The next three are capitals of states in central and western India where the rice–wheat cycle is not practised. The next two, Hyderabad and Bengaluru, are capitals in the middle of the Deccan Plateau of southern India, where dryland farming prevails over all but small riverine areas of the country. The last two, Mumbai and Chennai, are coastal cities situated on fertile but narrow coastal plains that receive high levels of rainfall and grow a variety of spices and cash crops in addition to some rice. In the data for October 2020, the AQI for PM2.5 had risen into the dark-orange zone because the non-Basmati rice harvest may already have begun by 5–6 October.

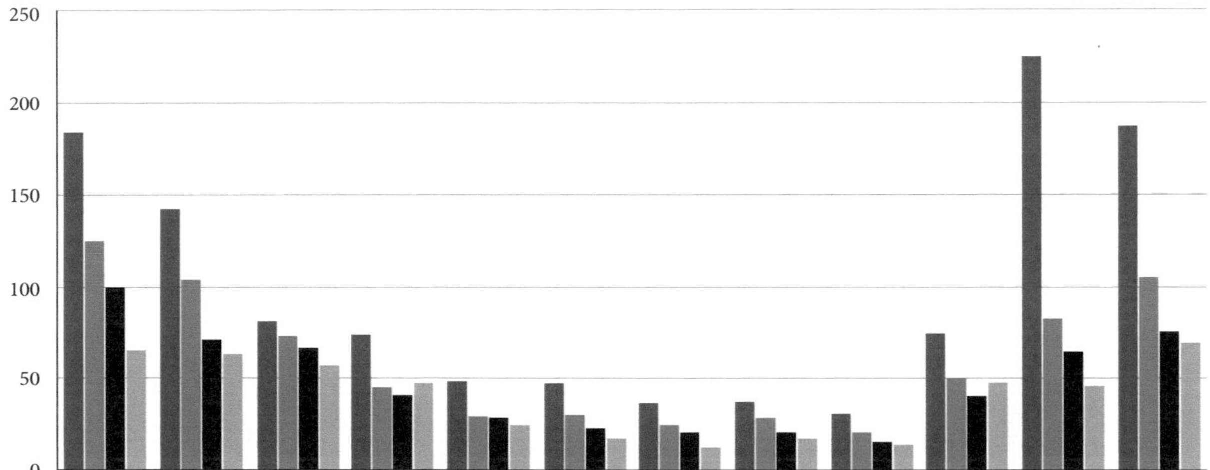

	1	2	3	4	5	6	7	8	9	10	11	12
Delhi	183.7	142.2	80.5	72.9	47.4	47.1	35.6	36.9	30.2	73.7	224.1	186.4
Kolkata	124.3	103.6	72.8	44.2	28.4	29.7	24	27.6	19.6	49.8	81.6	104.7
Mumbai	98.5	70.7	66	39.5	27.9	22.5	19.6	19.4	15.3	39.9	63.3	74.6
Hyderabad	64.6	63	56.4	46.9	23.6	16.9	12	16.7	13.6	46.7	45.5	68.4

Figure 6.6 Air Quality Index, 2021.

The first attempt to find a way to use the rice straw fruitfully was made by the Punjab state government in 2013 when it agreed to subsidize a small rural power plant to run on rice and wheat straw.[19] By 2019, nine power plants burning rice and wheat straw had been set up, with a total generating capacity of 98 MW. The average price at which the state power distribution company (DISCOM) was selling power to its customers was 6.5 cents (Rs 4.52)/kWh. But the cost of generating power from rice and wheat straw amounted to 10.3 cents (Rs 7.94/kWh). The difference of 3.8 cents (Rs 3.42) per unit made a state-wide replication of such plants unaffordable. The government of Punjab announced therefore that it would not sign any more such start-up contracts. It did not altogether rule out the use of rice straw as fuel but announced that the price would be determined through reverse bidding, as was being done for solar energy. In 2020, the first such auction fetched a bid of 7.7 cents (Rs 5.62) per unit. This was less than it cost industrial units to generate power from the diesel generators nearly all were using to supplement the perennially unreliable supply from the state grid. So, while use of straw and particularly rice-husks for power generation may continue to increase in the future, such plants will absorb at best a tiny fraction of the 35–40 Mt of rice straw and surplus wheat straw that Punjab generates every year.

By then, encouraged by the central government, Punjab and adjoining Haryana and western Uttar Pradesh had begun to look for other ways to eliminate straw burning. In April 2018, after the first Great Smog episode, the central government had launched a scheme entitled 'Promotion of agricultural mechanization for in situ management of crop residue in the states of Punjab, Haryana, Uttar Pradesh and NCT of Delhi'. Its purpose was to subsidize the purchase of an array of machines for in-situ crop residue management. It offered a 50% subsidy on such purchases to individual farmers and 80% to Cooperative Societies, Farmers' Interest Groups, and other collectives. The scheme also aimed to provide an assured market for such machinery, attracting investment for its manufacture and financing demonstrations on farmers' fields. The funds for it, Rs 11.5 billion ($155 million), were provided by the central government.

Since 85% of operational holdings in Punjab are less than 4 hectares in size, and even smaller in Haryana and west Uttar Pradesh, in effect the grants were intended to help establish Farm Machinery Banks and facilitate the deployment of crop residue management machinery, such as tractors, trailers, harvesters, and balers of straw. Heavy subsidies were offered for the purchase of a tiller-cum-seeder machine, promptly nicknamed the 'Happy Seeder', that breaks up the caked topsoil created by the drying of the flooded paddy fields, sows the wheat seeds, and applies the required dose of fertilizers without having to remove the remaining stubble. But as often happens in programmes that are drawn up in a hurry, neither the Punjab nor the Haryana government realized that the 'Happy Seeder' required a 50-horsepower tractor to pull it. In 2018, fewer than 10% of Punjab and Haryana's farmers had tractors with that horsepower.[20]

Exploration of other uses

The government began therefore to explore other options. One that gained some traction was to 'torrefy' (desiccate and briquette) the rice straw at a low temperature and sell it to power plants as a substitute for coal. Two power plants of the state-owned National Thermal Power Corporation (NTPC), located in Haryana south of Delhi, and one at Bhatinda in Punjab agreed to buy straw either as pellets or as torrefied briquettes and to blend these with coal for up to 7% of the fuel.[21] The NTPC baulked, however, at raising the ratio to 10% or higher without first studying the impact this would have on the heat generated for the boilers.

A more serious hurdle proved to be the price NTPC was willing to pay for the biomass briquettes, around $74 (Rs 5,500) per tonne for the normal and $87 (Rs 6,500) per tonne for the torrefied pellets.[22] This was about what they were paying for coal, but this did not cover the high cost of transporting the biomass over the fairly long distances to the power stations. As a result, beyond the initial commitments made by these power plants, the project did not get off the ground. The experiment did show, however, that torrefied straw could replace coal in small, decentralized power plants set up in rural areas.

The NITI-CII trials

In November 2018, when the second 'Great Smog' closed more than 1,800 primary schools in the four-state region for the first time, the National

Institution for Transforming India (NITI) Aayog joined hands with the Confederation of Indian Industry (CII) to 'design actionable steps' to address air pollution at its source, i.e. in the farmers' fields. Earlier in the year, CII had presented NITI with a 'Crop residue management project to curb the practice of crop residue burning in the north-western states'. Based on solutions identified by CII-NITI in a Biomass Management Report, CII set up a pilot project designed to judge farmers' willingness to take up sustainable agricultural practices in the four-state region. The project began in September 2018 and covered 6,400 hectares of farmland, owned or tilled by 3,000 farmers living in 19 villages of two districts in Punjab (Ludhiana and Patiala).Its goal was to persuade the 3,000 farmers in the trial to adopt ecologically friendly ways of using the paddy stubble instead of burning it, and to assess how it affected their farming, their health, and their incomes. The project was an unqualified success. Paddy straw and stubble burning ended in 74% of the 6,400 hectares. In the previous year the area spared had been 3%. Instead of burning 27,000 tonnes of straw and stubble as they used to, the farmers spread 25,000 tonnes on their fields as mulch and ploughed the remaining 2,000 tonnes back into the soil. The mulch increased soil moisture by 13% by preventing evaporation; and the ploughing back recycled into the soil a variety of nutrients valued at Rs 410 per tonne of straw and reduced the need for fertilizers for the rice–wheat cycle as a whole by 29%.

A significantly larger saving came from not having to use pesticides to kill the weeds that sprout in the fields alongside the wheat crop when the fields have been entirely cleared of paddy stubble by burning.[23] This was because the mulch prevented sunlight from reaching the weeds alongside the rows of wheat. The task force estimated that this saved farmers between Rs 800 and Rs 2,400 ($25.50 and $76.50) per hectare.

Perhaps best of all, without the straw burning the topsoil and forming a hard crust, the farmers did not have to plough and water their land repeatedly to make the germination of the wheat crop possible. This saved labour, time, and water. The project's agronomists estimated that there had been a saving of 2.5 billion litres of water over the 4,800 hectares on which the 'residual' straw and stubble had not been burned. Mulching and ploughing back the stubble had saved also between 4 and 10 days of labour.

Since these benefits are significant and every attempt was made to publicize them, why did paddy straw burning not diminish in the ensuing

3 years? The most probable reason, indeed the chief drawback of all solutions based on better soil management alone, is that most of the benefits come not in cash but in kind. Not only are these less palpable, but realizing them requires disrupting a network of economic relationships that stretches from farm labourers to the balers of straw, to a growing number of small- and medium-sized enterprises ranging from the power plants to manufacturers of packaging materials and paper, a resource supplied mainly by the balers.

In rural societies, these economic linkages are reinforced by familial, caste, clan, and other social relationships that people are unwilling to disrupt. Sensing this, in 2019 the state government of Punjab offered farmers a cash benefit if they did not burn their stubble, but then they reneged on this commitment, citing a lack of funds. When an appeal against this landed before the Supreme Court of India, the court directed all state governments to pay Rs 6,000 per hectare ($80) to every farmer who did not burn their stubble. But state governments ignored the directive. So the stubble burning continued unchecked in 2020 and 2021.[24] Between 15 September and 10 October 2021, Punjab recorded 764 incidents of crop residue burning. Of this number, 701 took place in just 4 days after the government declared the monsoon rains to have ended.[25]

In 2019, the central government conceded that paddy straw and stubble burning would stop only when a market could be found for this material that would give farmers a second assured source of income. So NITI Aayog and CII framed an ambitious plan for setting up 2,938 plants involving an investment of $6.2 billion (Rs 470 billion) to produce saleable products from rice and wheat straw, cattle dung, and other agro-residues in four states of India where the rice–wheat crop cycle was most in vogue. These were Punjab, Haryana, Uttar Pradesh, and Rajasthan. After intensive discussions with various industry organizations and rural credit institutions, they focused on captive power generation, paper-making, and bio-compressed natural gas (bio-CNG).

Captive power

As power shortages are not likely to disappear in the near future, the central government is encouraging captive power generation by industry using rice and wheat straw both unprocessed and torrefied into briquettes, or pyrolised into biochar. If this proves a success, it will remove the principal hurdle to the use of crop residues as fuel for power generation—the prohibitive cost of transporting biomass beyond very short distances. But as of September 2021, a market for torrefied briquettes had yet to develop. To encourage this, the government had asked the state-owned NTPC to place advertisements announcing that it was prepared to purchase torrefied briquets to replace coal in its thermal power plants. NTPC did so, but the price it offered, Rs 6,000–7,000 ($80–93) per tonne, was far too low: to produce a tonne of biochar briquettes would require three or more tonnes of straw. Since captive power plants were already buying rice straw at Rs 2,300 or thereabouts per tonne, not surprisingly NTPC received no response to its tender notice. The offer was apparently not serious and had been made only to satisfy the government, for gasifying a tonne of rice straw would yield at most a third of a tonne of biochar. Agronomists in Punjab concluded therefore that NTPC had only made the offer to comply with the central government's directive.

Paper

A second avenue for use of rice straw could be to make paper. In 2015, Trident Ltd, a $1 billion multi-product company that has specialized in manufacturing high-quality paper from pulp made from wheat straw, announced that it intended to double its captive power generation and meet its growing requirement for process heat using rice straw and husk as fuel. Trident is the largest Indian, and possibly the world's largest, producer of specialty papers from wheat straw. Its example could be followed by other paper makers in the four northern rice and wheat growing states.

Bio-CNG

The most recent and potentially most profitable initiative that Delhi and the governments of the four states have taken is to revive a programme that was first launched in the 1950s. This was to set up biogas plants in each village capable of digesting a mixture of straw and cattle dung and methane for use as a superior cooking fuel, available further to run irrigation pumps when the need arose. The scheme failed partly because the infrastructure needed to collect cattle dung and organic refuse in the villages did not exist and partly because the large difference between day and night and summer and winter temperatures seriously affected the fermentation process.

The current scheme is a vastly more complicated version of the original idea. Pioneered in Punjab by Verbio Vereinigte BioEnergie AG, a German manufacturer and supplier of biofuels headquartered in Zörbig, Saxony-Anhalt, it has specialized in producing biodiesel, ethanol fuel, and biogas on an industrial scale and has developed its own processes for so doing. Verbio's plant started production in December 2021. It has sealed digesters with a 5,200-cubic metre holding capacity and an elaborate system for maintaining a constant temperature and pressure. It is designed to ferment 300 tonnes of paddy straw mixed with 40 tonnes of cattle dung per day, assisted by enzymes identified by the company, and to produce 80,000 cubic metres (600 kg) of bio-CNG a day. Verbio's plant started delivering Gas in October 2022, but at the time of writing, the company had not revealed its finances. However, its capital cost was estimated at Rs 300 million ($4 million).

By mid-2021, the Punjab government had sanctioned 22 more plants of this design. Together these are expected to consume 731,400 tonnes of rice straw every year. Under a central government scheme for producing cellulosic ethanol, the state-owned Indian Oil Corporation is also setting up a plant that will consume 200,000 tonnes of rice straw per annum. These plants are expected to come on stream by the end of 2023.

Verbio's plant has drawn an enthusiastic response from the central government, prompting the government to decide that this will be its chosen weapon for implementing the NITI-CII plan to set up 2,938 enterprises based on rice straw to greatly limit, if not end, the menace of crop residue burning. By September 2021, it had announced its intention to issue Letters of Intent to produce bio-CNG and fertilizer from Verbio-type bio-digesters, and issued 1,500 of them, mostly to the state-owned oil companies that had earlier taken on the responsibility to produce cellulosic ethanol.

A second, highly desirable feature of the Verbio digester is its need for 40 tonnes of animal dung per day. This amounts to 13,200 tonnes a year on a 330-day working basis. Punjab is one of the centres of a dairying revolution that has made India the second largest exporter of dairy products in the world. Today, its dairy cattle produce 70.8 Mt of fresh dung, equivalent to 17.7 Mt of dry mass, per year. Should the Verbio digester prove financially viable, it will eliminate a potent source of methane emissions in India, the source from microbial processing of animal waste, while replacing fossil-based natural gas.

Crop residue gasification

The only route to eliminating paddy straw burning that neither the central nor any state government has even considered so far is one that could eliminate not only the life-threatening smog that now wraps Delhi and every other city in north and central India in winter, but also the mountains of garbage that have grown up around every city and the liquid wastes that are choking its drains and poisoning its drinking water. This is biomass gasification. Biomass gasification is as old as the making of charcoal. At the advent of the automobile age, it was used to produce motor fuels from coal and municipal waste, until it was displaced by oil. As will be described later in Chapter 12, this technology has now been refined to the point where it can produce every transport fuel, from aviation turbine fuel to pure 'green' hydrogen. From the point of view of India and every other developing country, the biomass gasification option has one other inestimable virtue: this is its capacity to create a self-sustaining, employment-intensive pattern of growth to replace the city-centred, fossil fuel-dependent model of growth that societies have been attempting to master since the end of the colonial age. This could slow urbanization and limit the further growth of existing megalopoli such as Delhi and Mumbai. Biomass gasification offers a potentially viable way to dispose also of the millions of tonnes of plastic that are fouling India's drainage systems, its rivers, and the oceans that surround it. These external 'social' benefits for India are discussed in greater detail in Chapters 12 and 15.

Notes and References

1. Basmati rice is the fragrant, extra-long-grain rice that has become a highly valued export of South Asia. It is a long-stemmed plant and takes 140 days to mature, as against the 93–110 days taken by short-stemmed high-yielding strains of non-basmati rice. As a result, only fields planted with the latter are double cropped with wheat. In Punjab and Haryana 70% of the total area under rice is planted with non-basmati rice. Basmati rice is too delicate to be harvested mechanically and is therefore harvested manually.
2. Rizwan, S.A., Nongkynrih, Baridalyne, and Kumar Gupta, Sanjeev (2013). Air pollution in Delhi: its magnitude and effects on health. *Indian Journal of Community Medicine* 38(1): January–March. https://www.ncbi.nlm.nih.gov/pmc/articles/PMC3612296/ (viewed 3/10/2021). In the early 1990s, the contribution of vehicles was especially high because nearly all two-wheelers and three-wheelers ran on two-stroke engines that burned a mixture of gasoline and oil, and lead was still being used as the anti-knock agent in gasoline for automobiles. By 2010, both these sources had been eliminated and all but a few of the cars with engines designed in the 1950s had gone off the road. But the air quality continued to worsen.
3. Gardiner Harris, *The New York Times*, 14 February 2015. https://www.nytimes.com/2015/02/15/world/asia/delhi-wakes-up-to-an-air-pollution-problem-it-cannot-ignore.html(viewed 17/3/2022).
4. Mukesh Sharma and Onkar Dikshit (2016): Comprehensive Study on Air Pollution and Green House Gases (GHGs) in Delhi (Final Report: Air Pollution Component). Submitted to Department of Environment, Government of National Capital Territory of Delhi, and Delhi Pollution Control Committee, Delhi. https://cerca.iitd.ac.in/uploads/Reports/1576211826iitk.pdf.
5. The seven main constituents of PM 2.5 are sulphates, nitrates, ammonium, elemental carbon, organic carbon, silicon, and sodium ion. PM10 consists mostly of compounds of the above that settle as dust along roadsides and construction sites, coarser particulate matter in smoke from fires, and compounds of the various chemicals that form the primary emissions.
6. Anubhuti Vishnoi, *The Economic Times*, 10 December 2015. https://economictimes.indiatimes.com/news/politics-and-nation/iit-kanpur-study-says-trucks-and-road-dust-are-bigger-pollutants-than-cars-in-delhi/articleshow/50115310.cms?utm_source=contentofinterest&utm_medium=text& utm_campaign=cppst.
7. Michael Safi: Indian government declares Delhi air pollution an emergency, *The Guardian*, 6 November 2016.
8. Delhi and NCR Air Pollution Report Card: 2017.
9. Reality Check, *BBC News*, 6 November 2019. https://www.bbc.com/news/world-asia-india-49729291 (viewed 18/3/2022).

10. Sunil Dhingra and Prem Shankar Jha: The Smog That Hides a Bonanza, *The Wire*, 29 November 2018.
11. Stubble burning emissions will double by 2050, if not controlled, says PGI study, *Hindustan Times*, 29 October 2018. https://www.hindustantimes.com/punjab/stubble-burning-emissions-will-double-by-2050-if-not-controlled-says-pgi-study/story-fW33nTBsBBEQtIqpstZNzJ.html (viewed 18/3/2022).
12. Swati Gupta and Sugam Pokharel, *CNN World*, 4 November 2019. https://www.cnn.com/2019/11/04/india/delhi-india-smog-pollution-intl-hnk/index.html (viewed 23/3/2022).
13. World Air Quality Report: 35 Indian cities in top 50 of world's most polluted list; here's how you can fight air pollution, Lifestyle Desk, *The Indian Express*, 23 March 2022. https://indianexpress.com/article/lifestyle/health/35-indian-cities-50-world-most-polluted-air-pollution-quality-aqi-pm-levels-7832341/ (viewed 23/3/2022).
14. Centre for Science and Environment: Delhi and NCR Air Pollution Report Card: 2017–18, 28 February 2018. https://www.cseindia.org/delhi-ncr-air-pollution-reportcard-2017-18-8521 (viewed 24/3/2022).
15. IQAir: The world's most polluted countries in 2020—PM 2.5 ranking. https://www.iqair.com/us/world-most-polluted-countries.
16. Centre for Science and Environment: EPCA releases Report Card on Delhi's air pollution, 28 February 2018. https://www.cseindia.org/epca-releases-report-card-on-delhi-s-air-pollution-8502 (viewed 24/3/2022).
17. India.com, 9 November 2020. https://www.india.com/news/india/air-quality-severe-in-delhi-ncr-aqi-pollution-imd-forecast-4203275/18. *Hindustan Times*, 21 September 2021: 'Data of the past 10 days showed that about 57% of Delhi's electricity supply is sourced from coal-fired plants, 18% is hydropower, 13% from gas-based stations, 10.2% from solar energy, 1.6% is wind energy and the balance (0.2%) is from waste-to-energy (WTE) plants in the city. To be sure, electricity demand is dynamic and the percentages can marginally vary depending on weather conditions which trigger power demand in the city.' https://www.hindustantimes.com/analysis/climate-how-delhi-can-transition-to-cleaner-green-power-101632219971991.html(viewed 24/3/2022).
18. Jasjeev Gandhiok: Just 8% of Delhi power procured from unclean source: CSE, *The Times of India*, 6 January 2021. http://timesofindia.indiatimes.com/articleshow/80131515.cms?utm_source=contentofinterest&utm_medium=text&utm_campaign=cppst (viewed 25/3/2022).
19. Amy Yee, *The New York Times*, 8 October 2013. https://www.nytimes.com/2013/10/09/business/energy-environment/india-increases-effort-to-harness-biomass-energy.html.
20. Sunil Dhingra and Prem Shankar Jha: The Smog That Hides a Bonanza, *The Wire*, 29 November 2018. https://science.thewire.in/author/sdhinga-psjha/ (viewed 25/3/2022). There are two ways to overcome this problem: the first is to make all farmers buy or rent the heavier tractors. The second is to replace the

relatively simple 'square balers' in common use, which cost around Rs 1.6 million ($22,000) each, with rotary balers that remove all but an inch or two of the stubble but cost Rs 28–35 lakh each. Some of these had appeared in Punjab, but few farm machine leasing companies were prepared to buy them because of the farmers' reluctance to pay the higher rental fee without any commensurate return.

21. Siddhesh Hande: Can biomass pellets reduce emissions in coal plants? *Down to Earth Magazine*, 25 August 2018. www.downtoearth.org.in (viewed 12/10/2021).
22. NTPC opens tenders for clean fuel suppliers to major TTP. www.renewablesnow.com.
23. Pesticides have to be sprayed three to four times in a season to kill a prevalent weed named *Phalaris minor* (Gullidanda) that grows alongside a wheat crop.
24. Stubble burning: why it continues to smother north India, *BBC News*, 30 November 2020. https://www.bbc.com/news/world-asia-india-54930380 (viewed 30/3/2022).
25. IARI data shows spike in farm fires post monsoon in Punjab, Haryana; situation better than last year, *Hindustan Times*, 12 October 2021. https://www.hindustantimes.com/cities/chandigarh-news/iari-data-shows-spike-in-farm-fires-post-monsoon-in-punjab-haryana-situation-better-than-last-year-101633980489073.html (viewed 30/3/2022).

7

India's Current Energy Consumption and Projected Demand

Like the nations of Europe, India's modern industrialization was based on coal. Coal was found in abundance in the country—first in Bihar and Bengal in the mid-19th century, then half a century later in the north-east, then clubbed together in the single giant province of Assam, and a little later in central India. But unlike Europe, which began to shift its primary energy base from coal to oil shortly after the end of the Second World War, India has continued to rely on coal to meet the vast majority of its energy needs up to the present day.

All it has done in the past hundred or more years is to supplement coal with oil and natural gas. In 2019, India's primary energy demand amounted to 929 million tonnes of oil equivalent (Mtoe). Figure 7.1 provides a breakdown of the individual contributions to this total, with sources to the left and applications to the right. The sources are identified, in order of Mtoe, as coal 413 Mtoe, oil 243 Mtoe, traditional biomass 113 Mtoe, modern renewables 95 Mtoe, natural gas 55 Mtoe, and nuclear 10 Mtoe. Expressed as percentages of the total, coal was responsible for 44.5%, oil 26.2%, traditional biomass 12.2%, modern renewables 10.2%, natural gas 5.9%, and nuclear 1.1%. The category labelled as modern renewables includes the contribution from large-scale hydro, responsible for generation of 175 TWh of electricity in 2019, equivalent to 15.1 Mtoe. Wind and solar in combination contributed an additional 114 TWh to power generation in 2019, equivalent to 9.8 Mtoe.

Of the total of 929 Mtoe of primary energy developed in 2019, 284 Mtoe (30.1%) was lost during the conversion to useful energy. Of the remaining 645 Mtoe, industry consumed 269 Mtoe (42%), transport 108 Mtoe (17%), and buildings 217 Mtoe (34%). The balance goes into agriculture and miscellaneous household uses in the villages. Fossil sources

Sun, Wind, and Biomass. Prem Shankar Jha and Michael B. McElroy, Oxford University Press.
 DOI: 10.1093/9780198944768.003.0007

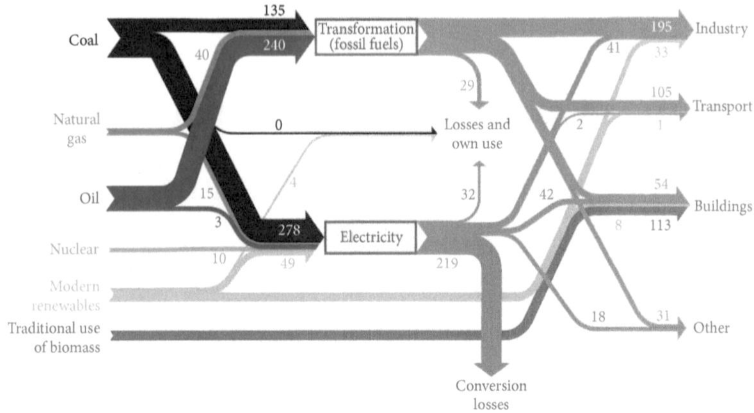

Figure 7.1 Flows of energy in India's energy system, 2019 (in mtoe).
Source: IEA (2021), India Energy Outlook 2021, https://www.iea.org/reports/india-energy-outlook-2021.

accounted for 77% of the country's primary energy supply in 2019. The supply from modern renewables, including hydro, wind, and solar, consumed in industry and buildings amounted to just a little more than 10%, emphasizing the magnitude of the challenge India faces as it seeks to meet the goal for carbon neutrality by 2070 announced by Prime Minister Narendra Modi at the Glasgow Summit in November 2021. The breakdown in CO_2 emissions, summarized for 2019 in Figure 7.2, provides context for Modi's commitment.

So great is the challenge that up to days before the Glasgow Summit, members of Modi's cabinet were still insisting that the burden of reducing greenhouse gas emissions should be apportioned on the basis of individual countries' *cumulative*, and not *current* contributions to the increase in the CO_2 content of the atmosphere.

It is not surprising therefore that, although India began to explore the potential for renewable energy as far back as 2009 when it established its first Solar Energy Mission, as recently as of now no study had been implemented to explore what would be required to shift out of fossil fuels altogether. So when Modi announced India's commitment, he threw out a challenge not only to the rest of the world but also to his own countrymen.

India has been slow to recognize and even slower to accept the challenge posed by the need to arrest climate change. In the run-up to the

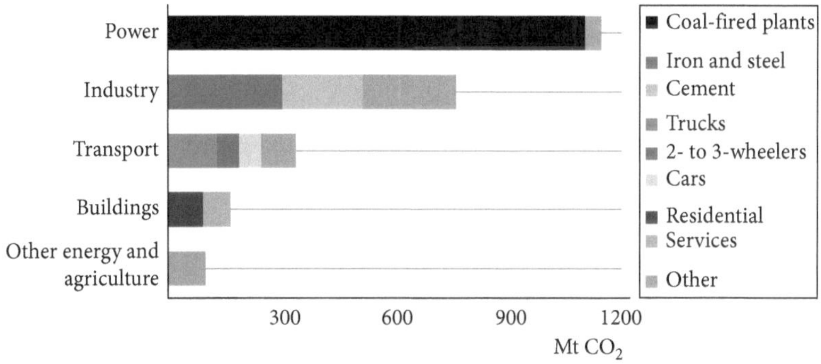

Figure 7.2 CO_2 emissions from the Indian energy sector, 2019.
Source: IEA (2021), India Energy Outlook 2021, https://www.iea.org/reports/india-energy-outlook-2021.

Paris Climate Summit in November 2015, the IPCC had warned that to have a reasonable chance of meeting the 2 °C target the world would have to limit the addition of CO_2 to the atmosphere to at most 1,000 Gt by 2100. But even if all countries met their nationally determined commitments (NDCs), the investments already announced by fossil fuel producers and consumers would add 723 Gt of CO_2 to the atmosphere by 2030. To stay within the 1,000-Gt limit therefore, the world would have to bring down annual CO_2 emissions from 40 Gt a year to 0 in 14 years.[1] This was manifestly impossible.

But in 2015, New Delhi had not yet got the message for, during the Paris conference, it stubbornly opposed bringing down the target limit on global warming by 2100 from 2 °C to 1.5 °C, and eventually had its way. In the closing stages of the conference, Prakash Javadekar, India's Minister for the Environment, expressed his satisfaction that, 'India has been able to secure its interest in the agreement'. After the conference ended, Prime Minister Narendra Modi claimed, with obvious relief, that 'climate justice has won'.[2] Days later, India's Minister for Coal and Power, Piyush Goyal, reiterated that India intended to double its coal output by 2020. It is therefore hardly surprising that India came to be regarded as a spoiler at the conference.

From 1994, when the UN Framework Commission for Climate Change (UNFCCC) drafted the Kyoto Protocol, up to the Copenhagen Summit of 2009, India had stuck firmly to the belief that, since the problem had been created by the industrialized countries, it was theirs to solve. In 1994, all parties to the UNFCCC had agreed to exempt the developing countries from the burden of reducing fossil fuel emissions. But as the pace of economic growth accelerated dramatically in Asia, Eastern Europe, and South America in the 1990s and 2000s and developing country emissions first equalled and then exceeded those of the industrialized 'North', the Conference of Parties to the UNFCCC (commonly referred to as the COP) could no longer delude itself that this was a tenable position.

India's intransigence was understandable: while its per capita carbon 'footprint' was still only 2.1 tonnes a year, among the lowest in the world, with more than 1.1 billion inhabitants, its *total emissions* were the third largest in the world. What is more, from 1993 to 2012, with a brief dip at the turn of the millennium, it had been recording an average GDP growth of more than 7% a year, and doing this with a relatively modest annual increase in its consumption of energy, 5.1% per year.

The official stance of the government remained therefore adamantly opposed to any dilution, much less scrapping, of the Kyoto Protocol which was adopted by COP3 in Kyoto in December 1997 and entered into force on 16 February 2005. It changed only in the last hours of the Copenhagen conference in 2009, after a 3-hour meeting between President Barack Obama and the heads of governments of China, India, Brazil, and South Africa. The change followed a *volte face* by China, whose CO_2 emissions had exceeded those of the US for the first time in 2008. So, especially for China, the developing country exemption no longer made sense.

The agreement that emerged from Obama's discussion with the leaders in Copenhagen led to replacement of the Kyoto Protocol with what came to be known as the Copenhagen Accord. Under this, all of the COP signatories would set independently determined targets for emissions reduction, subjected to collective scrutiny at subsequent Summit meetings.

India accepted this change. The United Progressive Alliance government then in power set up committees to assess various modes of carbon capture and sequestration and to examine alternative, renewable energy technologies that could be employed to replace fossil fuels. To do the latter, it set up a test station for wind turbine power in 2011 and the first National Solar Energy Mission to assess the cost and reliability of solar thermal and solar photovoltaic power generation.

But most of the government's attention remained focused on adaptation to climate change, and on ways to minimize the impacts of climate change on the poor in particular. Despite the urgent warnings of the IPCC in its Fifth Assessment Report, which was released in 2014, energy planning within the Indian government remained focused on (1) *moderating* the demand that would result from continued high economic growth by increasing energy efficiency; (2) *minimizing* threats to India's energy security; and (3) *ensuring* access to clean energy for the poor. Reducing India's overall carbon footprint remained a distant objective—a goal towards which India would move, perhaps spontaneously, as the cost of renewable energy continued to decline.

In October 2015, 2 months before the Paris Summit, India submitted three commitments to the UNFCCC: (1) by 2030 it would reduce the emissions intensity per dollar of GDP of its economy by 33% to 35% below the 2005 level; (2) by 2022, 40% of its electricity-generating capacity would be based on non-fossil energy sources; and (3) it would increase the area under forest and tree cover to create an additional annual carbon sink of 2.5–3 billion tonnes of CO_2.[3]

Although hailed by the international press, these were not ambitious commitments. Thanks mainly to a sharp slowdown in the rate of industrial growth after 2010, a succession of technological improvements such as a spontaneous shift from less to more efficient automobiles and applications, and the replacement of tungsten light bulbs with LED bulbs, India's emissions per rupee of GDP had already declined by 42% from 0.465 megajoules (MJ) per rupee in 2006–07 to 0.271 MJ per rupee in 2014–15.[4] Thus, at the Paris conference, India's representatives felt no qualms about playing the spoiler's role and opposing the setting of more stringent emission reduction goals, such as the 1.5 °C limit on global warming favoured by scientists and smaller nation members of the COP. In fact, as mentioned already, 2 days after the Summit ended, India's energy minister, Piyush Goyal, reiterated that India would increase its coal-fired generating capacity from 170,000 MW in 2015 to 568,000 MW in 2030. He stated also that wind, solar, and 'biogas' energy would bring down the country's dependence on coal for power generation only from 61% to 57% over the following 15 years!

This pessimism about the ability of renewable energy technologies to replace fossil fuels had been fuelled by two projections of India's energy needs, which were released, one just before the other, shortly after the Paris Summit. The first was *The Draft National Energy Policy* of the Narendra Modi government, released by NITI Aayog, India's rechristened Planning Commission, before India had decided on its NDC for the Paris Summit. An updated version of this report based on the assumption of a long-term average growth of GDP of 8% was released in 2017. The second was a policy paper titled '*A more sustainable energy strategy for India*', written by Montek Ahluwalia, Himanshu Gupta, and Nicholas Stern, released 6 months after the Summit, in July 2016. This covered a longer time horizon, from 2012 to 2047, and estimated the overall increase and changing pattern of India's energy system over this period.[5]

Structural change in the economy

The goal of both papers was to assess the increase in demand for energy that would result from rapid growth in the economy over the next 25–35 years and the impact this would have on the country's dependence on imported coal, oil, and gas. Their secondary purpose was to assess by how much India's carbon footprint could be reduced through increases in energy efficiency combined with a shift to renewable energy. Both papers used a forecasting tool developed by the Planning Commission in 2012–13, under the auspices of Ahluwalia (who was then its Deputy Chairman), called the *Indian Energy Security Scenario (IESS)*. The IESS allowed researchers to project future energy demand for various rates of GDP growth and structural change over eight sectors of the economy identified as residential and commercial buildings (i.e. the two construction sectors), passenger and freight transport, industry, agriculture, cooking, and telecommunications.

The National Energy Policy (NEP) updated its 2015 paper and assumed a growth rate of 8%. For this it sketched out both a 'business-as-usual' and an 'ambitious' scenario for the period 2014–2040. The Ahluwalia–Gupta–Stern study assumed a rate of growth of 7.4%. With this growth rate the IESS projected that:

(1) per capita GDP would rise 8.7 times from $1440 in 2012 to $12,583 in 2047;

(2) the urban population would increase from 36% to 52% of the total;
(3) per capita transport requirements would rise by 3.1 times for passengers and 6.3 times for freight; and
(4) the residential space requirement would increase by a factor of three for every person. The consumption of cement would consequently rise from 190 kg to 611 kg per person per year; demand for steel from 66 kg to 384 kg; and car ownership from 9 per 1,000 persons to 70 per 1,000 persons.[6]

Projected energy consumption

The Ahluwalia–Gupta–Stern paper, which constructed a 'business-as-usual' and a low carbon future model, added one more variable to its analysis. This was the *level of effort* that the government was prepared to make to achieve a low carbon future. Based on extensive consultations with energy experts, the authors concluded that it was possible to make a more dramatic improvement in energy efficiency than the NEP had anticipated, and that it could achieve a reduction of as much as 86% in the 35 years of the model.[7] This assumed an annual rise in energy efficiency of 1.79%, considerably higher than the efficiency of 1.58% that applied between 2006 and 2021. In contrast to the NEP, they were relatively pessimistic about the possibility of replacing fossil fuels with renewable energy and concluded that this would meet only 14% of the demand for energy in 2047. All in all, they concluded that while GDP would grow by 12 times, the use of fossil fuels would rise by as little as a factor of only 2.6. India's carbon footprint would rise therefore from 1.7 tonnes of CO_2 emissions per capita in 2012 to 3.0 tonnes in 2047, less than the 3.8 tonnes estimated under their 'business-as-usual' scenario.

Arresting climate change may not provide sufficient motivation

A common feature of both estimates was their insularity. Neither showed any awareness that climate change was not only accelerating, but also approaching several potential 'tipping points'. Both were perfectly comfortable with the idea that India would have done enough if it only limited the *growth* of its CO_2 emissions by the middle of the century.

Neither then nor after the IPCC issued its dire warning in the 1.5 °C Special Report of 2018 did either study even remotely try to model what would be needed to bring emissions down to net zero by the end of the century, let alone by 2050 or 2060 that the IPCC's Special Report had strongly recommended.

In 2015, this pessimism had still been forgivable because renewable energy—even wind power—was still more expensive than coal or natural gas. It was a settled belief then that renewable energy would remain more expensive than fossil fuels for several more years, if not decades. So it is not surprising that the Indian government did not explore the possibility that replacing fossil fuel with renewable energy technologies might actually sustain, and possibly even accelerate, economic growth. The government did not, therefore, feel any qualms in predicting that to sustain an 8% rate of GDP growth, India would still be using coal to meet its need for base load power, and therefore that its coal-fired power generating capacity would have to grow from 125 GW in 2012 to 441 GW in 2040.

The only energy-related goal on which both the Modi NEP and the Ahluwalia–Gupta–Stern paper agreed was the need to minimize India's dependence on imported oil and coal. But even in its low carbon scenario, the latter paper estimated that India's dependence on imported coal would rise from 18% in 2012 to 57% in 2047, on oil from 77% to 90%, and on imported natural gas from 22% to 43%.

Extreme weather events—beginnings of change

So rapid has been the change of perceptions since then that, in 2023, a mere 7 years after these reports were published, both appear archaic and irrelevant. The change has been brought about by a succession of extreme weather events and a near simultaneous, dramatic fall in the cost of wind and solar energy generation that began at about the time of the Paris Summit.

The first extreme weather event that caught the entire nation's attention had occurred 10 years earlier, in July 2005, when the port city of Mumbai received 994 mm (*nearly 40 inches*)—of rain in 24 hours. Of this deluge, 709 mm (28 inches) fell in just 5 hours, between 2.30 and 7.30 in the afternoon. The floods that swept the city claimed 1,493 lives.[8] This was treated as a freak occurrence, but as similar, though less disastrous, deluges followed and cyclones began to develop in the Arabian Sea, where they had not been known before, it was recognized, in retrospect, as a harbinger of the destabilization of the climate system that was already underway. From 2015 on, the number of such events multiplied both in frequency and intensity.

In November 2015, Chennai, the 3,000-year-old port city of Tamil Nadu on the east coast, was visited by its first flood in more than a hundred years. This was because the heaviest rains in several decades had forced the state government to release 30,000 cubic feet per second (cusecs) of water from the Chembarambakkam reservoir into the Adyar River that flows through Chennai. This made the river overflow its banks and submerge neighbourhoods on both sides, taking 421 lives and causing more than $3 billion of damage.[9] The government had been left with no alternative but to open the floodgates at Chembarambakkam because the district of Neyveli, in which the dam is located, had received 483 mm of rain (more than 17 inches) within 2 days. Similar deluges had lashed the districts of Cuddalore, Chidambaram, and Chennai, further augmenting the inflow to the reservoir. That this was not another freak occurrence became apparent when a similar, though somewhat less severe, flood ravaged Chennai in the same month of 2017 only 2 years later.[10] On that occasion, Chennai received more than 40% of its average November rainfall *in a single day* on 3 November.

The summer of 2018, the very next year, saw an even worse catastrophe: late in the monsoons, an unprecedented deluge engulfed Kerala, India's southernmost state, most of which lies along the western edge of the Deccan Plateau. Kerala is where the annual south-west monsoons make their landfall, typically at the end of May. The 2018 monsoon had been exceptionally good, so the state had already received 75% more than its normal quota of rain by August. When the deluge occurred on 8 August, the reservoirs of many of its dams were already overflowing. The state authorities were forced therefore to open the gates of the barrages to relieve the pressure. The resulting floods swept away homes, roads, railway tracks, and power lines in 14 districts, killed 483 people, and caused $5.6 billion worth of damage.[11]

When weather extremes become more marked, droughts often follow floods. The most severe and prolonged drought in living memory occurred the very next year, in 2018–19, when the north-east monsoon, which should have hit the east coast of southern India in October, pretty much failed. As a result, by the time the next south-west monsoon arrived in June 2019, virtually the entire Deccan Plateau had been in the grip of the worst drought in living memory for 10 months. To make matters worse for the Deccan, the arrival of the monsoon was severely delayed. In June, the country received only 33% of its normal quota of rain. So, for many parts of the Deccan, the drought extended all the way to July.

The 2019 experience highlighted also another feature of the contemporary monsoon—its increasing unpredictability. For despite the deficit in June, the overall precipitation turned out to be 10% above the long-term average. This was because the monsoons continued well into September, accompanied by 52% more precipitation than the normal average.

One of the effects of this growing unpredictability was a rise in the frequency of cloudbursts and thus severe flooding in several parts of the country, notably central India, the northern state of Uttar Pradesh, and the eastern state of Bihar. The worst affected state in 2019 was Bihar, where flash floods ravaged 13 out of 38 districts, more than a third of the state, and left 47 persons and thousands of head of cattle dead in their wake. This succession of catastrophes made the Indian Meteorological Office concede that extreme weather events had become a regular feature of the climate and that it was being caused by global warming.[12]

In 2019, the Arabian Sea experienced three tropical cyclones, and on 20 May, in the early summer of 2020, the mangrove-forested region of the Sunderbans, in Bengal, and adjoining Bangladesh were hit by Amphan, the first super cyclone in 21 years, which created tidal waves as high as 5 m (15 ft) and inundated low-lying areas of the islands in both countries (Figure 7.3).

Figure 7.3 Satellite picture of super cyclone Amphan.

Source: The National Aeronautics and Space Administration (NASA)—EOSDIS Worldview (https://worldview.earthdata.nasa.gov/), Public Domain, https://commons.wikimedia.org/w/index.php?curid=90414562.

India's first report on the impact of global warming

On 16 June 2020, the Ministry of Earth Sciences released its first ever climate change assessment by government scientific institutions. Its predictions were grim: by the end of this century the average temperature in India was projected to rise by 4.4 °C. Heat waves would become three to four times more frequent, the intensity of tropical cyclones would increase dramatically, and the average sea level would rise by 30 cm above what it had been in the three decades from 1986 to 2015.[13] In these decades, the average temperature on the warmest days and coldest nights of the year had risen by 0.63 °C and 0.40 °C respectively. If the world does nothing to limit global warming (the 'business-as-usual' scenario) these temperatures are likely to rise by another 4.7 °C and 5.5 °C respectively by the end of the century. There would be 55–70% more warm days and warm nights than there had been between 1976 and 2005. Heat waves would be three to four times more frequent and would last almost twice as long. The report was unequivocal in its finding that this rapid change in India's climate would put severe stress on the country's biodiversity, agricultural output, and freshwater resources, and increase the risk of heat strokes, cardiovascular and neurological diseases, and other stress-related disorders. The number of warm days has already increased by about 9.9 per decade and warm nights by 7.7 per decade, the report said. Cold nights have decreased by 3.3 per decade during this period. Warm days or nights are identified with times when the maximum and minimum temperatures are above the 90th percentile.

The report placed the blame for the intensifying cycle of floods and drought on *an abnormal increase in sea surface temperature over the Indian Ocean, by more than 1 degree in the 64 years between 1951 and 2015*, half as much more than the average increase in global average sea surface warming of 0.7 °C. The frequency of extremely severe cyclonic storms (ESCS) over the Arabian Sea increased during the post-monsoon seasons of 1998–2018. Most models are projecting a higher sea surface warming in the Arabian Sea than the Bay of Bengal. Concentrations of oxygen and marine phytoplankton have declined in recent decades.

The report underlined that climate change has already skewed the monsoon patterns, with rains decreasing by 6% between 1951 and 2015. It confirmed the surmise that increases in aerosol emissions associated with burning of agricultural wastes are likely to have contributed to this decline. The Indo-Gangetic Plains and Western Ghats recorded the highest declines, but the frequency of extreme rainfall (over 150 mm) increased by 75% between 1950 and 2015 in central India. There is a shift towards more frequent dry spells (27% higher than the 1951–1980 period) and more intense wet spells during the summer monsoon season. High elevations of the Tibetan Plateau have recorded severe warming, as high as 0.5 °C per decade, according to the report. The report warned also of a cascade of climate-related hazards that would overlap. For instance, a region could experience an abnormally long or intense summer heat wave followed by intense monsoon floods alternating with lengthening dry spells. 'Low-lying coastal zones, especially on India's east coast, may witness rising sea levels damaging property and increasing groundwater salinity. A rise in cyclone intensities will likely result in increasing inundation from the accompanying storm surges that turn proximate agricultural lands and lakes saline, and imperil wildlife', the report added.[14]

Wind and solar photovoltaic power

Since then, the government has given its full support to increasing the share of renewable energy in electricity generation. Prime Minister Narendra Modi, who had set the ball rolling shortly after he came to power in 2014 by setting a target of 100,000 MW of solar and wind power generation capacity by 2019, raised the target to 175,000 MW by 2022 and to 275,000 MW by 2027. According to recent calculations, if the present tempo of the increase in renewable energy investment is maintained, generating capacity could reach 470,000 MW by 2030.

But the impact of these policy decisions is being vitiated by a continuing inertia in the government's policies. This is most noticeable in an asymmetry in the subsidies for energy development that has severely discouraged investment. This was highlighted in 2016 by the International Institute for Sustainable Development (IISD), which noted that while subsidies for renewables in India had increased from $431 million in 2014 to $1.4 billion in 2016, subsidies for electricity, more than four-fifths of which was still being generated from coal and gas, *were around six times this amount.*[15] Three years later, in 2019, the subsidies for consumption of electricity these had increased dramatically to $15.6 billion (Rs 1.103 trillion), or 0.52% of GDP.[16]

Notes and References

1. Paris, 2015: Climate Agreement. www.Carbon-Price.com.
2. PM Narendra Modi on Paris deal: climate justice won, *The Times of India*, 13 December 2015. https://timesofindia.indiatimes.com/india/pm-narendra-modi-on-paris-deal-climate-justice-won/articleshow/50159914.cms.
3. India's Intended Nationally Determined Contribution: Working Towards Climate Justice. https://www4.unfccc.int/sites/submissions/INDC/Published%20Documents/India/1/INDIA%20INDC%20TO%20UNFCCC.pdf (viewed 4/4/2022).
4. TERI, 2015, cited by K. Sreelatha Reddy and Jyothi Mehra in 'Energy intensity in India and its impact on GDP', *IOSR Journal of Business and Management*, pp. 6–12. http://www.iosrjournals.org/iosr-jbm/papers/Conf.17037-2017/Volume-9/2.%2006-12.pdf (viewed 4/4/2022).
5. Prepared under the auspices of the Indian Council for Research on International Economic Relations (ICRIER) and the Grantham Research Institute of the London School of Economics.
6. Ahluwalia–Gupta–Stern paper, Appendix 1.
7. See Table 7: Sources of Reduction in Emissions from BAU to LC in 2047.
8. It's official: Maharashtra flood toll is 1493, *rediff News*, 29 August 2005. https://www.rediff.com/news/2005/aug/29rain.htm (viewed 5/4/2022).
9. 2015 Chennai Floods Which Took 421 Lives Was a 'Man-Made Disaster': CAG, 2018. '300 or more lives' https://www.ndtv.com/chennai-news/2015-chennai-floods-a-man-made-disaster-says-cag-report-1881906 (viewed 5/4/2022).
10. India—Torrential Rain Causes Deadly Flooding in Tamil Nadu, FloodList, 3 November 2017. https://floodlist.com/asia/india-floods-tamil-nadu-november-2017 (viewed 6/4/2022).
11. 2018 Kerala floods. https://en.wikipedia.org/wiki/2018_Kerala_floods (viewed 7/4/2022).
12. Srishti Choudhary: Global warming is behind extreme rains: IMD chief, 15 October 2019. https://www.livemint.com/news/india/global-warming-is-behind-extreme-rains-imd-chief-11571157861949.html (viewed 13/4/2022).
13. Assessment of Climate Change over the Indian Region: A Report of the Ministry of Earth Sciences (MoES), Government of India. https://link.springer.com/content/pdf/10.1007%2F978-981-15-4327-2.pdf (viewed 14/4/2022).
14. Ibid.
15. https://www.iisd.org/system/files/publications/stories-g20-india-en.pdf.
16. IISD Webinar: Better targeting of residential electricity subsidies, 16 October 2020. https://www.iisd.org/gsi/news-events/webinar-better-targeting-residential-electricity-subsidies(viewed 08/01/2025).

8

Where Is India on the Road to Renewables?

India's commitment to shifting out of fossil fuels by 2070 may bring strategizing within the government into sharp focus. This is essential because, so far, except in the areas of wind and SPV power generation, that focus has been notable for its absence. The country took its first steps to developing renewable energy when Prime Minister Manmohan Singh launched the country's first *National Solar Mission*, named after India's first Prime Minister, Jawaharlal Nehru on 11 January 2010. This was an exploratory mission designed to test various parameters on which future plans could be based. It therefore set a very modest target of 20 GW of solar-generating capacity to be attained by 2022. Half of this was to be met by SPV grid-connected power plants and rooftop installations and the other half by concentrated solar thermal (CSP). To manage the Mission, the government set up in 2011 a new parastatal company, the Solar Energy Corporation of India.

The National Solar Mission opted for equal investments in SPV and CSP generating capacity, reflecting the fact that SPV power in 2010 was slightly more expensive than CSP. The disparity, although small, was apparent in the first auctions, when, against a feed-in tariff of Rs 12.16 per kWh awarded for the first SPV plant in 2010, the first 50-megawatt (MW) CSP plant, set up at Anantapur in an arid district of Tamil Nadu in South India, proposed to generate 110 GWh of power and to supply it to the state grid at a rate of Rs 11.31 per kWh.

But in the next 7 years, while the cost of SPV power fell rapidly, a series of teething troubles in the first two large CSP plants that came online in India made both miss their power generation targets and fail to break-even at the committed power purchase prices. This and the simultaneous development of unfavourable financial conditions created by a

Sun, Wind, and Biomass. Prem Shankar Jha and Michael B. McElroy, Oxford University Press.
 DOI: 10.1093/9780198944768.003.0008

sharp rise in interest rates after 2011 caused a rapid loss of interest in CSP power generation. By contrast, SPV systems remained free from teething troubles, and a rapid fall in the cost of silicon cells brought the capital cost of SPV systems down from $3.50 per watt of generating capacity in 2013 to just over $2.00 per watt in 2017. As a result, while investment in SPV plants raced ahead, CSP lagged behind and by 2019 had no takers in the Indian market.

The rapid growth of SPV power

The main cause of SPV's success was the dramatic fall in the price of silicon cells, as indicated in Figure 8.1, from close to $2.00 per watt in 2010 to $0.30 per watt in 2015,[1] driven in large measure by the entry of Chinese firms into the world market. As a consequence, in a succession of 'reverse' auctions (many bidders, one buyer) the average bid price fell from Rs 12.16 per kWh (25 US cents)[2] in 2010 to Rs 3.15 (4.63 cents) in April 2017 and to Rs 2.71 (without incentives) in May 2018[3] and finally to Rs 1.99 in December 2020.

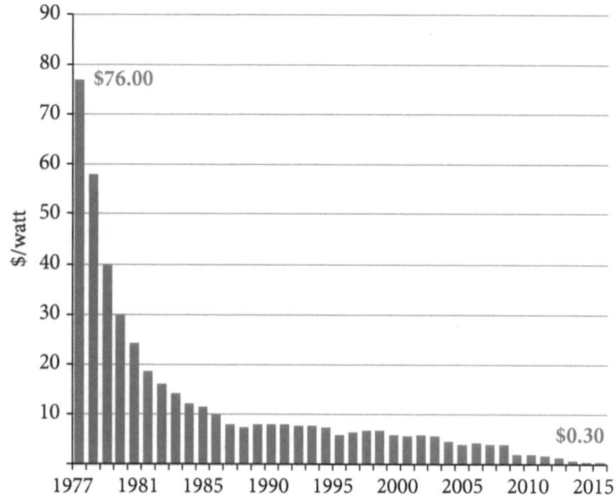

Figure 8.1 Price of silicon photovoltaic cells, 1977–2015.

Source: Rfassbind, https://commons.wikimedia.org/wiki/File:Price_history_of_silicon_PV_cells_since_1977.svg, based on Bloomberg New Energy Finance and www.pv.energytrend.com.

Table 8.1 Installed SPV on 31 March, 2010–2019.

Year	Cumulative capacity (GW)	Electricity generation (TWh)	Capacity factor (%)	Hours a year*
2010	0.161	–		
2011	0.461	–		
2012	1.205	–		
2013	2.319	–		
2014	2.63	3.35	14.5	–
2015	3.744	4.60	15.6	1368
2016	6.763	7.45	16.6	1452
2017	12.289	12.09	16.2	1420
2018	21.651	25.87	14.5	1269
2019	28.181	39.27	18.0	1577
Average			16.2	1417

* Calculated by dividing the power generated in a given year by the generation capacity installed in the previous year plus half the capacity added in the current year.

Source: Prepared by authors from current data.

The accompanying Table 8.1 shows that in 6 years, from 2013–14 to 2018–19, grid-connected SPV plants generated power for an average of 1,417.2 hours a year, equivalent to a capacity factor (hours of power supplied to the grid in a year as a percentage of the total number of daylight hours) of 16.2%.[4]

The construction cost of SPV power plants has not fallen as rapidly globally as the decline in the cost of silicon panels. So the cost of the panels has had a progressively smaller effect on the total system cost of PV utilities. By contrast, in India, as illustrated in Figure 8.2, where construction costs are a much smaller fraction of the total system cost, the decline in the overall cost of SPV generation has been more pronounced and has gone on longer. As a result, in 2019, the investment required in India per watt of installed capacity was a third of what it was in Japan, half of that in Germany, 40% below that in the US, and, surprisingly, 18% below that of China. India has benefited therefore the most from the steep fall in solar panel prices.

Not surprisingly, the growth of installed SPV capacity has been impressive in India. It grew from 0.16 GW in 2009–10 to 30.87 GW as of 30

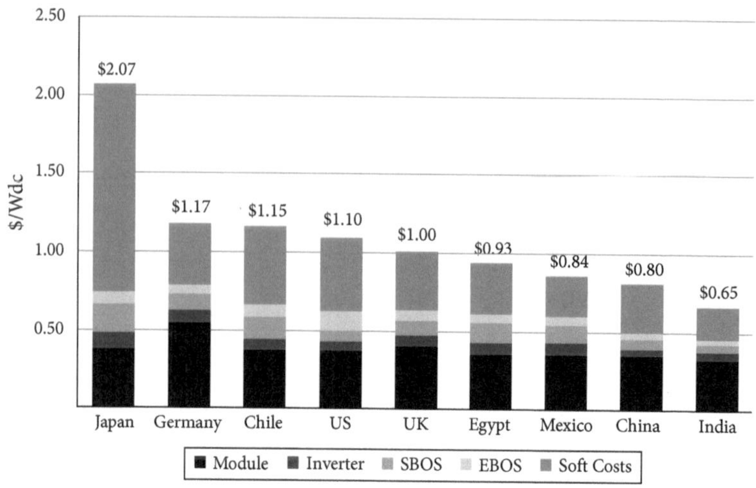

Figure 8.2 International 2017 utility PV system pricing by country [cost per watt of direct current (Wdc), in $].
Source: IRENA's 'Renewable power generation costs in 2018' report (May 2019).

September 2019. The onset of the COVID-19 pandemic brought the rate of expansion down sharply to 3.2 GW in calendar year 2020, but this was followed by a rebound to 10 GW in 2021–22.[5] India added 13.5 GW of renewable energy generation capacity in fiscal year 2021–22. This included 12.4 GW of solar and 1.1 GW of wind power capacity. This was more than double the capacity added in the previous fiscal year.

By May 2022, India's installed solar power generation capacity had risen to 52 GW.[6] This was greatly in excess of the National Solar Mission's original target of 20 GW of capacity by 2022, but significantly short of the more ambitious target of 100 GW that Prime Minister Narendra Modi had set in 2017. However, the lag involves barely 2 years since there were over 54 GW of utility-scale projects in the pipeline at various stages of development, and another 33 GW of projects had been tendered and were awaiting auction.[7]

Why has CSP failed in India?

The National Solar Mission sanctioned the construction of ten CSP plants between 2008 and 2014. Three of these investments involved pilot-scale plants designed to gain familiarity with the different technologies then in use. The first was a 1-MW plant, set up in 2008 to gain familiarity with the parabolic trough system which was then the standard technology in use in Spain and elsewhere in the world. The second was a 2.5-MW plant set up by a private firm, Acme Engineering, in Rajasthan in 2011. This was designed to explore the then emerging power tower technology. The third was a 14-MW CSP plant set up by the state-owned National Thermal Power Corporation in 2017 alongside a large combination gas/coal-fired conventional power plant at Dadri, on the outskirts of New Delhi. This investment was designed to meet some of the plant's demand for electricity while reducing the need for deployment of expensive, largely imported, natural gas. This plant has been an unqualified success but, somewhat surprisingly, has not so far been duplicated elsewhere. Of the seven remaining plants, only three were still in operation in 2022: two are located in Rajasthan, one in Tamil Nadu. An interesting fact is that although the storage of heat as molten salt had been developed several years earlier, none of these three plants had been equipped with this facility at the time of writing.

As of 2024, there were 112 operational CSP plants across the world with a total capacity installed capacity of 6.7 Gigawatts with an additional 7 under construction. Of the first 60, 24 were equipped to accommodate storage of heat, with potentials for storage ranging from 2.5 to 15 hours. *All but two of these plants were sanctioned after 2012, when round the clock power from solar thermal plants became possible.* These began coming into operation on or after 2014.[8] Despite its high levels of insolation India has lagged far behind, preferring the easy, short incubation and cheaper option of solar photo-voltaic power.

In 2020 the total capacity of the operational Concentrating Solar Power (CSP) plants in the world amounted to 6.6 GW. India accounted for a mere 0.23 GW of this. In sharp contrast, the installed capacity of Solar PV installations amounted to 940 GW in 2021 52 GW of this was in India. In 2020, therefore, CSP still accounted for a minor portion of the world's capacity to generate solar energy. The discrepancy in India was even greater. While CSP accounted for (0.7%), of worldwide solar power generation capacity, it accounted for only 0.4% in India. The cause of the discrepancy between investments in CSP versus SPV was clearly the much higher initial capital cost of generating power from the former.

In 2017–18 this was double or more of the cost of SPV power. The chief drawback of SPV, however was its periodicity and dependence upon the weather, In May 2018, therefore, the US Department of Energy began a research programme designed to bring the cost of CSP power to $0.05 per kWh.[9] This has matured and changed the future prospects for CSP in the US. It has begun to do so in India too. In the first quarter of 2024, India put out tenders for renewable energy projects, that required over 50% to be supplied by Concentrated Solar Power (CSP). Three such projects have been selected:

NTPC has invited an Expression of Interest (EOI) for setting up Concentrated Solar Power (CSP) for the following cases on Build Own Operate (B-O-O) basis.

1: Inter State Transmission System (ISTS) 50 MW RE power supply from Concentrated Solar Power (CSP) Project with Thermal Energy Storage System (TESS) for 8 hours operation during peak hours & non-solar hours.

2: An ISTS connected 100 MW (net at ex-bus) RE-RTC power supply for peak hours/non-solar hours for 8 hours duration using only wind power and CSP along with dedicated thermal energy storage system for CSP of minimum capacity 200 MWh (electrical). The applicant shall have the option to sell the surplus power.

3: ISTS connected 100 MW (net at ex-bus) RE-RTC power supply for 24 hours duration using CSP along with dedicated thermal energy storage system for CSP of minimum capacity 200 MWh (electrical) and other RE source like Solar PV, Wind etc. The required energy storage for RE source (solar/wind) shall also be included. The applicant shall have the option to sell the surplus power.

Wind power generation

Wind power generation has been the other success story in India. What is more, several studies, including two by colleagues in the School of Engineering and Applied Sciences at Harvard University, described in detail in the following chapters, show that India's wind energy potential is somewhat greater than that of solar Photovoltaic power. But despite this, there has been an inexplicable slowdown in the growth of wind investments since 2019–20, predating the onset of the COVID-19 pandemic.

Wind mapping was begun by the government in 1960, organized through its recently established National Aeronautics Laboratory. Following 25 years of data collection, the first demonstration projects using 55-KW wind turbines were set up in the coastal areas of Maharashtra and Gujarat on the west coast of India and in Tamil Nadu on the east coast. The first comprehensive estimate of wind power potential was made by the National Institute for Wind Energy (NIWE) in 2010. It rated this at 40.13 GW for wind turbines operating at hub heights of 50 m above the ground. In a subsequent estimate for turbines placed 80 m above the ground, it raised the estimate to 102.13 GW. In a third still more detailed estimate, in 2015, with the help of data on land use and land cover made available by the National Remote Sensing Centre (NSRC) and terrain elevation and wind density data obtained from other international remote sensing sources, it raised the estimate to 302.25 GW at 100-m turbine height.[10]

More recent estimates of India's wind power generation capacity, presented in the following chapters, show that even NIWE's 2015 estimate is only a fraction of the existing national wind power potential. The reason for the vast disparity is that NIWE had been extremely conservative in its estimate of the total land that would be available for wind farming, setting this at only 2% of the land area of the country, i.e. 65,745 km^2. On this basis, it had further excluded land that was being used or would be needed for economic development such as property in cities and villages, roads, railways, airports, water bodies, areas with elevation of more than 1,500 m, where the air gets too thin to turn the wind turbines efficiently, and slopes that exceeded 20 degrees. It excluded also land over which the wind was not sufficiently strong or consistently capable of accommodating a Capacity Utilisation Factor (commonly referred to as the Capacity Factor) of more than 20%.

NIWE then grouped the remaining land under three levels of suitability—wasteland, cultivable land, and forest land—and assumed that 80% of the wasteland, 30% of the cultivable land, and 5% of the forest land would be available for siting wind farms. The more rigorous exclusion of unsuitable land in the third estimate reduced the available land to 50,375 km^2 but increased the wind power potential to 302 GW. It arrived at this estimate by assuming that three turbines with a capacity of 2 MW could be installed on every square kilometre of the available land.[11] This study also placed them 100 and 120 m above the ground, and, like NIWE's third estimate, carefully excluded all land that is urban, covered by water bodies, forests, plantations, orchards, mountains, snow, or ice, and large buffer zones around all villages, urban areas, industry, and locations of important infrastructure. It also excluded land more than 1,500 m above sea level and areas without wind sufficient to accommodate an average capacity utilisation factor of 15%.[12] With all these exclusions, they estimated that the land area that could be brought under onshore wind farm development amounted to 947,983 km^2 for wind turbines at a 100-m hub height and 1,114,182 km^2 for turbines at 120 m.

The authors presented two estimates of the generation potential from this land. The first was for turbines placed 5 by 7 blade diameters apart on the two axes. The second was for turbines separated by 3 by 5 diameters.[13] The former configuration would enable each square kilometre to accommodate 5.7 MW of generating capacity. The latter would raise the generating capacity per square kilometre to 13.3 MW. The authors concluded that, using 40% of the available land, the former configuration would enable India to generate 2,162 GW with turbines at 100 m height and 2,541 GW with turbines at 120 m. The latter configuration would raise the generation capacity to 5,043 GW at 100-m turbine height and 5,927 GW at 120 m. But this configuration is likely to create considerable interference between the turbines and thus reduce their efficiency.

Policy hurdles set up by competing parastatal organisations have slowed growth in solar power generation

The following estimates serve to highlight the long road India has to travel. So far, although overseen and coordinated by the Solar Energy Corporation of India, wind power generation has been an entirely private initiative. Helped initially by highly supportive government policies, generation capacity grew rapidly to 23 GW in April 2015 (Table 8.2). Then, despite Prime Minister Narendra Modi setting a target of 60 GW

Table 8.2 History of wind development in terms of generation.

Year	2013–14	2014–15	2015–16	2016–17	2017–18	2018–19
Generation Capacity (MW)	22,465	23,447	26,777	32,280	34,046	36,625
Generation (GWh)	–	28,214	28,604	46,011	52,666	62,036

Source: Prepared by authors from current data.

for 2019, the pace of growth slowed significantly. By September 2019, generating capacity had risen to only 36.6 GW. Since then, growth has slowed further. Only 1.1 GW of generating capacity was added in 2021–22. At the end of March 2022, the total wind power generation capacity had risen to only 40.2 GW.[14]

There could be more than one reason for the slowdown, which has been historically notable for the solar option and which began long before the COVID-19 pandemic hit the country. One possible explanation is that reverse auctions inherently involve high-risk ventures for investors. Competition among bidders pares profits to the bone and leaves the winners with little cushion against unforeseen price shocks and technological glitches. This has reduced the number of bidders at successive auctions. But a more important cause is the growing disregard that state governments are displaying for the sanctity of contracts in their obsession with securing the lowest bid possible for renewable energy projects. As the cost of solar panels has fallen, the bid prices in successive auctions have followed suit. This has made state governments with older purchasing power agreement (PPAs) renege on their contracts and don't force investors to renegotiate the prices.

The trend started in February 2021 when the Gujarat government reneged on the PPA that was signed for a 700-MW SPV plant in 2020 to sell power to the state grid at Rs 2.78–2.81 per kWh. The government tried to renegotiate a lower price. It was prompted to do this by its success in getting a bid of Rs 1.99 for another project 6 months after signing the first agreement.[15] The state High Court overturned that decision, but this did not deter other state governments from following Gujarat's example. Over the next few months, Andhra Pradesh, then Punjab, and most recently the government of Uttar Pradesh withheld issuing the Letter of

Intent (LOI) for more than a year after the reverse auction had been held. Since then, this trend has spread like a pandemic through the country.[16] In May 2021, the state of Uttar Pradesh cancelled the solar auction process for installing 500-MW solar power, hoping to get a lower price through a new auction. The original contract was signed for 500 MW of solar-generating facilities in February 2020. Four companies, including Saudi Arabia's Al Jomiah, won the auction quoting a price of Rs 3.17–3.18 (4.2–4.3 cents per kWh. But another auction held in November in Rajasthan and Gujarat yielded a price as low as Rs 1.99 (2.7 cents) per unit. The justification given by the director of the Uttar Pradesh New and Renewable Energy Development Agency (UPNEDA) was that: 'In the light of recent events, where very low solar tariffs have been discovered in Rajasthan and Gujarat and the bids invited by the Solar Energy Corporation of India, it seems very likely that a *fresh bid* may deliver better tariffs for consumers of Uttar Pradesh, in which case you are cordially invited to participate in the same'.[17] At present, Uttar Pradesh has an installed capacity of 3,878 MW of renewable energy, while the target set for the state by the Ministry of New and Renewable Energy was 14,221 MW by 2022, including 10,697 MW of solar power. The cancellation is expected to further slow Uttar Pradesh's progress. Other states such as Andhra Pradesh, Rajasthan, and Gujarat have taken similar steps with private developers. This trend of states trying to renegotiate a lower price for contracts already signed or commitments already made is a dangerous journey downhill that can destabilize the sector.

Subrahmanyam Pulipaka, Chief Executive Officer of the National Solar Energy Federation of India (NSEFI), which is India's umbrella organization of all solar energy stakeholders, said that the trend started when the Gujarat government tried to renegotiate the Purchase price agreement Post Acceptance, a decision overturned by the court. It spread then to the Andhra Pradesh government trying to renegotiate the deal, followed by Punjab requesting a discount on tariffs, Gujarat cancelling LoIs, and now Uttar Pradesh cancelling a tender after the reverse auction, without issuing the LoIs for more than a year. This trend of renegotiation of contracts has spread like a pandemic and industry's interests have been constantly overlooked. There is much therefore that needs to be set right in the government's policies if it truly wants to accelerate the development of solar and wind energy.

Another reason for the slowdown could be a growing reluctance of landowners to part with their land. This is not a significant hurdle for wind power generation where the need for an optimum dispersal of wind turbines ensures that only 2–3% at most of the land is taken out of its current use. But SPV plants occupy all of the land on which they are built, leaving the owner with little or no availability for other purposes. If this land is leased or purchased, it creates no problem. But in India, following a British practice that began in 1868, land is acquired for public purposes by the government at prices that it decides. The owners do not have the right of refusal and cannot therefore contest the acquisition, or the compensation being offered, except in a court of law.

The experience of India's first solar park, at Charanka in Gujarat, offers a glimpse into why a continuation of the present uncoordinated and unregulated policies of the state governments, with no overall guidance from the central government, could lead to a drying up of the supply of land for SPV farms. In 2010, the state government acquired an unspecified amount of privately owned grazing land in order to set up the country's first solar park, as depicted in Figure 8.3. Ten years later, 36 companies had set up SPV plants on 2,178 hectares of land with a generating capacity of 730 MW of power. But the people whom the government dispossessed have been left insecure and impoverished. Ten years after the project began, the villagers of Charanka are still waiting for the clean drinking water, free electricity, and irrigation they had been promised. Against a promise of 1,000 permanent jobs, only 60 people in the village were hired for washing panels and as security guards and grass cutters. The project created no jobs for women. This brought families who did not have land and their sons close to destitution and made them the worst victims of the solar park.[18]

A study released by the Delhi-based Observer Research Foundation in May 2022 concluded that while the area's most favourable for solar radiation throughout the year coincide with India's wastelands, most projections of solar energy development locate only 11–12% of it in deserts and dry scrublands. Wasteland is not favoured by project developers because locating there increases costs, partly because of the inhospitable terrain, partly because of the lack of supporting infrastructure, and partly because of the power transmission infrastructure that needs to be built to connect these remote areas to the grid. All of these factors act to increase

(a)

(b)

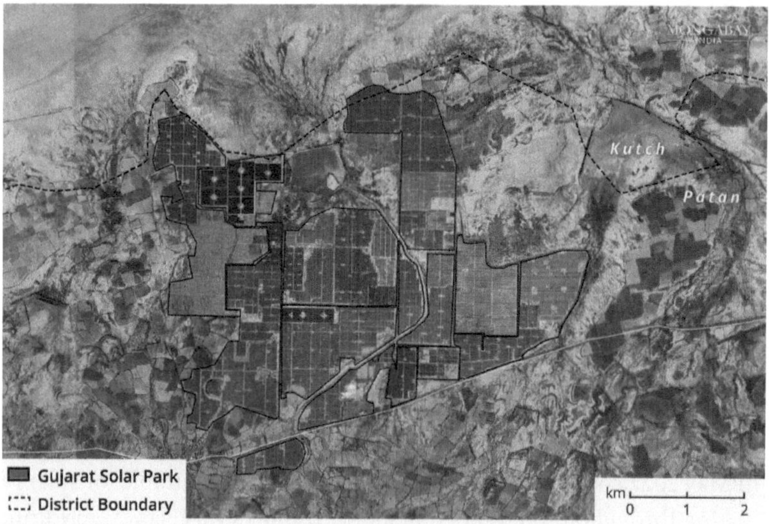

Figure 8.3 Ground (a) and satellite (b) views of Charanka Park.
Source: Google Earth, OpenStreetMap; map by www.techforwildlife.com/March 2022.

the cost of the project. The study concluded therefore that project developers preferred land that was under cultivation. Since then, in 2015–16 over 68% of land holdings were vested in marginal farmers who owned less than 1 hectare of land, millions of poor landowners and their rights have been treated as collateral damage in a mission for a cleaner and better world.[19]

The Indian government has been attempting to boost the solar sector in its pursuit of ambitious renewable energy targets. However, the growth of this sector was seriously hampered by the flip-flops on policy by state governments which, as discussed, had been repeatedly renegotiating contracts with power developers to make them lower their price per unit of electricity all the way till 2022.

As a result India fell substantially short of the target it had set, of installing 175 GW of solar power generation capacity by 2022, by achieving only 102.01 MW of solar power generating capacity by March 2023.[20] The increasing trend of state governments trying to cancel or renegotiate the contracts including PPAs that were finalized after a proper bidding process had added to the woes of the solar sector. Industry experts observe that this trend has the potential to derail India's long-term journey towards the green economy in that it could erode the confidence of investors.

Transport fuels

To date, India has made little progress with the development of renewable energy to meet the need for transport fuels and process heat for industry, both of which require major investments of energy. At the time of writing, the government has no clearly defined strategy to achieve the shift to renewables in these areas, and has adopted a scatter-gun approach of trying everything and waiting to see which technologies hit the mark.

In marked contrast to the strategy adopted by Dr Manmohan Singh's government in 2010 of placing the responsibility for proving a technology on the shoulders of the government and encouraging private investment by setting liberal feed-in tariffs, the Modi government set up no supporting parastatal enterprise to perform the nurturing function for start-ups in the transport and process heat sectors. As a result, both NITI Aayog and the Ministry of New and Renewable Energy adopted a 'follow the leader' approach to the greening of transport fuels and process heat and took their cues from technology developments in the industrialized countries. The first area in which it did this was the promotion of ethanol, initially as an additive and then as a possible replacement for gasoline. The second was the replacement of liquid-fuelled vehicles by electrically propelled alternatives (electric vehicles, EVs).

Between 2016 and 2018, NITI Aayog presented three proposals for the development of alternatives to oil and natural gas as transport fuels: ethanol to blend with gasoline; electricity-driven vehicles (EVs) to replace liquid-fuelled vehicles; and, most recently, the production of methanol and dimethyl-ether (DME) as alternatives to ethanol as 'green fuels', complemented further by liquefied petroleum gas produced by biomass gasification. By 2024, all this has succeeded in doing is to open the doors for sugar mills and dedicated biofuel plants to convert food crops—rice and maize and spoilt fruit—into ethanol at one extreme, and to promote the manufacture and sale of EVs, with handsome subsidies, on the other.

Chasing ethanol as a green fuel

In January 2013, the federal government launched an Ethanol Blended Petrol (EBP) Programme, which made it mandatory for oil companies to sell petrol blended with at least 5% ethanol. What is more, in 2018 the government announced a National Policy on Biofuels that set a target of 20% blending with ethanol by 2022. It did so in the full knowledge that while Brazilian sugar mills were making ethanol at the expense of sugar, Indian sugar mills were constrained from doing this because of the large domestic demand for sugar and were therefore making ethanol from the molasses left behind after making the sugar. This contained only 4.5% of the juice in the cane. It also knew that with many competing alternative uses for ethanol and simply not enough molasses to go around, the limit to blending was bound to be reached long before 20%, and even the intermediate target of 10% blend for green fuel would likely be unattainable.

Ethanol from molasses

In 2013, the Ministry of Petroleum and Natural Gas, which had been entrusted with the Ethanol blended Petrol program implementation, estimated that the sugar industry was producing on average around 3 billion litres (750 million US gallons) of ethanol a year. Of this, about 1.3 billion litres was going into making liquor. This was a sacred cow that no one could touch, not only because the liquor industry could always outbid the oil marketing companies, but also because the excise duty on liquor was responsible for the bulk of the self-generated tax revenues of the state governments.

Out of the remaining 1.7 billion litres, 0.6 to 0.8 billion litres was going into the manufacture of a variety of chemicals and pharmaceuticals, such as formaldehyde. That left a mere 1.1 to 0.9 billion litres for blending with gasoline. At the 2013–14 level of gasoline consumption, this was sufficient for blending only up to 3 percent of ethanol with the gasoline.

During the cane crushing season from 1 December 2015 to 30 November 2016, the oil marketing companies were able to purchase only 1.11 billion litres of ethanol from the sugar cane factories. In the very next year, a drought in Karnataka and Maharashtra, the two main sugar cane-producing states in southern India, reduced the output of sugar cane, and therefore sugar and ethanol, by a significant amount. As a result, the oil refineries were able to procure only 665 million litres of ethanol for blending with gasoline.

For 2017–18, the Indian Sugar Mills Association (ISMA) estimated that contracts could be signed to supply 1.4 billion litres to the oil refineries. This would have sufficed to raise the blending level to 4.22%.[21] But the sugar mills were able to supply 1.5 billion litres.[22] For the 2018–19 cane crushing season (18 December 2018 to 19 November 2019) therefore the ISMA signed contracts to supply 2.37 billion litres, which would suffice to raise the level of blending to 7.22%.

The rapid increase in contracted sales from 1.11 billion litres in 2016 to 2.37 billion litres in 2019 was made possible by a succession of increases in the mandated price of ethanol, from Rs 27 (about $0.59) per litre in 2011 to Rs 59.48 ($0.79) per litre in 2018. The rapid response of the sugar industry persuaded the government to raise its ethanol blending target to 10% by 2022 and to 20% by 2030. In 2022, the government advanced this 20% target date from 2030 to 2025.

The shift to food crops

To increase the ethanol content of gasoline to 10%, the oil marketing companies had estimated that the supply of ethanol would have to rise to 5.1 billion litres in 2021–22. But the COVID-19 pandemic and the change of work habits that followed reduced the requirement to 4.37 billion litres in 2021–22. The sugar industry met this comfortably. However, the consumption of gasoline has been growing at 9% a year,[23] almost twice as fast as the increase in consumption of crude oil, so the demand for gasoline could treble by 2030. Moving from a blending level of 7.22% to 20% will require therefore an eight-fold increase in the supply of ethanol, to about 18 billion litres.

The ISMA had been urging the government to allow the industry to increase its ethanol-making capacity.[24] It pointed out that sugar prices were stable and India had already attained parity with Brazil as an exporter in the world market. The government complied and raised the mandated purchase price of ethanol to enable sugar mills to increase their output at the expense of sugar.

The government's decision was contained in the National Policy on Biofuels announcement in 2018. The policy allowed the 'use of other than molasses as a feedstock for the manufacture of ethanol in the country. This includes sugarcane juice, damaged food grains, rotten potato, corn, and surplus food grains if any etc.'[25]

The policy statement sounds responsible, even humane, for post-harvest losses of fruits and vegetables alone in India are estimated to cost farmers Rs 133 billion ($1.7 billion) in lost revenues. But when put into practice it could easily have the opposite effect because the line between what is fit and unfit for human consumption will be set by the buyers and sellers themselves and will be impossible to police. In the end therefore it will be the market that will decide the price of the produce and the market will have a powerful new buyer—the biofuels industry—with whom the poor will not be able to compete. This has been a bone of contention between the Central government and the cereals farmers of northern India since 2020 and has yet to be resolved.

An idea of how powerful the pull of the transport sector will be may be had from the amount of starchy feedstock that will be needed to meet the 20% ethanol-blend target in 2030. One tonne of corn (maize), or any other grain or vegetable consisting of 70% starch, yields 420 litres of ethanol. Thus, to produce the additional 23.44 billion litres of ethanol, the country will have to divert 56 Mt of cereals or starchy vegetables from feeding human beings to feeding passenger transport vehicles.[26] And that would be only for starters, as the country would still be 80% short of the net zero emissions requirement to address the climate issue.

The hunt for cellulosic ethanol

In its National Policy on Biofuels of 2018, the government also decided to make a massive push to produce cellulosic ethanol from crop wastes. To do this it decided to set up twelve small commercial plants. In contrast to its policy on SPV plants, it did not hold any reverse auctions. Nor did it assign any of the twelve plants to the private sector. Instead, it delegated the task of setting them up to the three main government-owned oil marketing companies: the Indian Oil Corporation (IOCL), Hindustan Petroleum Corporation Limited (HPCL), and Bharat Petroleum Corporation Limited (BPCL). It tasked each of them with setting up four plants, each of which would process around 600 tonnes of biomass a day. These were estimated to cost $100–125 million each and would be subsidized to the extent of 15–20% of the capital cost.

A leap in the dark

The decision was not only a leap in the dark, but also the one taken without any prior study of the state of the technology, and without studying why, despite having been pursued for almost three decades, cellulosic ethanol had not become a viable technology anywhere in the world. In September 2019, the European Commission's Joint Research Centre had published a review of the status of cellulosic ethanol production titled 'What is still limiting the deployment of cellulosic ethanol? Analysis of the current status of the sector'. The study concluded that, 'despite the policy support (in terms of targets) and significant R&D funding in the last decade (both in EU and outside the EU), cellulosic ethanol production appears to be still limited. The review shows that the cellulosic ethanol sector appears to be still stagnating, characterized by technical difficulties as well as high production costs.'[27] Instead, the government launched a $1.5 billion plan based on assurances from two Indian companies claiming that they had developed, or had access to, proven technologies.

The oil marketing companies too, it seems, did no such research, for they invited tenders for twelve small commercial plants, designed to process 600 tonnes a day of biomass to be set up across eleven states of the country. Each was expected to cost between $110 million and $125 million and to produce between 149 and 181 litres of cellulosic ethanol per tonne of biomass. Ten of these twelve contracts were signed with the two Indian companies that claimed they had the technology. These were medium-sized companies with no prior experience of setting up or running a commercial cellulosic ethanol plant. The first was Pune-based Praj Industries, a company with a 2018–19 turnover of Rs 1,181 crores ($155 million), whose experience was largely confined to setting up molasses-based ethanol plants attached to sugar mills, and to the development of wastewater treatment plants. The company's credentials for going into completely unexplored territory rested solely on its successful demonstration of the production of ethanol from corn (maize) fibre.[28]

Corn fibre accounts for 10% of the dry weight of the maize and is a part of the residue, amounting to 23–24% by weight, that is left behind when it is liquefied and fermented. This residue, called distillers' dry goods with solubles (DDGS), is a protein-rich animal feed. But the presence of close to 40% of fibre in it greatly reduces its value. Since corn fibre contains a residual amount of starch, a technology had been developed in the US for extracting the fibre, softening it by exposing it to pressurized steam, grinding it into paste, and fermenting it. This increased the ethanol yield from the plant by up to 13%, and simultaneously enhanced the value of the DDGS as animal feed.

Since this process converts only the residual starch into ethanol, it yields only normal, and not cellulosic, ethanol. But Praj Industries claimed that the process could be modified to produce cellulosic ethanol.

In 2019, it announced its intention to set up a pilot plant to demonstrate this, but provided no details on the feedstock or the technology it intended to use Since March 2019, nothing more has been heard on the fate of its pilot plant. In that month, the government hosted a 3-day EU–India Conference on Advanced Biofuels, held in New Delhi.[29] But Praj Industries did not take part in this and provided no information on their technology for the booklet on advanced biochemical, thermochemical, and biogas technologies that the conference produced afterwards.[30] In spite of this, its technology was chosen by BPCL, one of the three oil majors, for a 480-tonne-per-day plant based on rice straw to be set up at Bargarh in Odisha. Praj Industries was retained also to provide the technology for four cellulosic ethanol plants in Punjab and Haryana.[31] As of June 2024, none of these plants had come on stream in either of the two states.

The second initiative was a technology developed by the DBT-ICT Centre for Energy Biosciences at the Institute of Chemical Technology, Mumbai.[32] In March 2016, it had started production of cellulosic ethanol in a pilot plant within the premises of India Glycols, a $680 million turnover company based in Uttarakhand that had established a solid reputation for production of glycols and high-quality (extra neutral) ethyl alcohol. By 2018, the plant had been run with different feedstocks for 4,000 hours. The DBT-ICT Centre convinced the Indian oil majors that its technology could be scaled up easily to 500 tonnes a day. So, it has been assigned two plants—at Bhatinda in Punjab and Bina in Madhya Pradesh. The foundation stone for the biorefinery in Bhatinda, Punjab was laid in 2018.

Among the twelve commercial-scale cellulosic ethanol plant that have been sanctioned so far the only one that has a good chance of succeeding is being built by Chempolis, a Finnish company that has entered India with the intention of building the first 'biorefinery' in the world to manufacture ethanol from bamboo. Chempolis is building a biorefinery scheduled to produce 50,000 tonnes (63.5 million litres) of ethanol from 1,000 tonnes per day of bamboo at Numaligarh in Assam, at a capital cost of 160 million Euros. This amounts to 181 litres per tonne of bamboo. Chempolis is using its own patented technology, *Formicobio*, and must be convinced that it will work with bamboo, given that it is prepared to make such a large investment. But the company had counted on being able to procure 300,000 tons of bamboo annually to produce 50,000 tons of bioethanol, and various byproducts, including 19,000 tons of furfural, 11,000 tons of acetic acid, and 144 gigawatt hours of green energy. It has found since then that it underestimated the difficulty of procuring such a large amount of bamboo every day, given its growing scarcity in India and the many other uses to which it is already being applied (notably scaffolding for construction sites). But Chempolis claimed that their technology would work equally well with straw, sugar cane bagasse, and other woody biomasses. So, it has modified the feedstock to a mixture of bamboo and rice straw. The plant is expected to start production in 2025. Its expected yield from this mixed feed is not known.

If Chempolis is able to keep its cost of production below the 2020 government-mandated price that the oil companies have to pay for ethanol, Rs 59.48 ($0. 80) per litre, it could well become the first company in the world to truly establish the viability of a large-scale second-generation bioethanol plant. But this does not, at first sight, seem likely.[33] If one assumes a plant life of 20 years, an interest rate of 5% on the capital, operation and maintenance costs of 2% of capital cost, and an overall return to the investors of 10%, then, at the October 2020 cost of Rs 6,000 per tonne for cut, green bamboo, plus a transport cost of Rs 1,000 per tonne, the annual cost will come to 54.6 million Euros (at 8 October 2020 exchange rate), while the annual revenue, at Rs 59.48 per litre and working for 350 days a year, will amount to 43.81 million Euros. For Chempolis to make a 10% overall rate of profit therefore, cellulosic ethanol will have to be priced at close to Rs 75 per litre. For it to just breakeven, the ethanol would still have to be priced at Rs 61 per litre. The economics of the remaining ten plants, which will be based on rice straw and bagasse, are unlikely to be any better.

In 2021, a study of the cost of producing cellulosic ethanol in India, commissioned by the International Council on Clean Technologies (ICCT), estimated that the cost of producing it without subsidization would be Rs 92–94 per litre ($1.10 in 2024) for the smaller plants producing 36.5 million litres a year and Rs 79–81 per litre ($0.95) for the larger plants producing 70 million litres a year. This was far higher than the government-mandated price for ethanol provided to the oil refineries, of Rs 62.65 ($0.85) per litre.[34]

A Faster Adoption and Manufacturing of Electric Vehicles (FAME) programme has been taken up by the Indian government

As with ethanol and cellulosic ethanol, India has also taken its cues from the West and decided to replace fossil fuels in the transport sector with electricity. In February 2017, NITI Aayog, India's renamed Planning Commission, hosted a 'multi-stakeholder' conference with the Rocky Mountain Institute and a selection of Indian industrialists to frame a strategy for replacing fossil fuels with electricity. In a 90-page report titled 'Transformative Mobility Solutions for India', they proposed a 15-year plan for shifting out of fossil fuels into EVs, claiming that it was based on policies similar to those that had been adopted in China. The plan involved limiting the registration of conventional vehicles through public lotteries and giving preferential treatment to registration of EVs. To kick-start the shift, the report suggested an initial bulk procurement of EVs, and the manufacture of standardized, exchangeable batteries for motorcycles, motor scooters, and auto rickshaws to bring down their cost. It also suggested setting 'favourable', i.e. subsidized, electricity tariffs for recharging electric cars. Three months after this meeting, in May 2017, NITI Aayog electrified the Indian elite (and created consternation in its car manufacturers) by releasing a report that proclaimed that India would replace its entire passenger vehicle fleet with electric cars by 2040. 'India can save 64% of anticipated passenger road-based mobility-related energy demand and 37% of carbon emissions in 2030 by pursuing a shared, electric, and connected mobility future', it announced. 'This would result in a reduction of 156 Mtoe (million tonnes of oil equivalent) in diesel and petrol consumption for that year.'[35] It did not elaborate on how it had arrived at these figures. In the same month, Piyush Goyal, the Minister of State for Power and Renewable Energy, announced that not a single petrol or diesel passenger vehicle would be sold in India from 2030 onwards. 'The Central government', he said, was 'preparing a road map to ensure that only electric vehicles would be produced and sold in the country by 2030. Different ministries had given their inputs and ideas, as part of a national mission to promote solar energy power plants and electric vehicles'.[36] Goyal's next statement revealed the will-o-the=wisp that was luring the government towards

EVs: 'The shift from petrol and diesel fuelled vehicles will' he said reduce India's dependence on oil and the cost of imports. Besides, it will strengthen the rupee and the current account deficit [in India's balance of payments] will disappear.'

The government he pointed out, was looking for simple solutions to a multiplicity of problems. The most important of these was India's almost total dependence on imported oil. Goyal's concern was entirely understandable. In 2017, 85% of all of the oil India consumed was imported and this figure was rising by the year. EVS, he believed, would greatly reduce But, as with its jump into Cellulosic Ethanol, the government's understanding of the technology shift that was involved was seriously deficient, for the Minister also claimed that while 'electric vehicles would not be cheaper to buy, they would more than make up for this by being much cheaper to run *and* the operating costs would also reduce', when all that manufacturers over the world are claiming is that the lower running cost of EVs would offset the higher capital cost of the vehicle and make EVs the cheaper option in the long run.

Automobile manufacturers were quick to point out the wishful thinking that supported this view. In particular, Maruti Udyog, in which the Indian government is a major partner with Suzuki Motors of Japan, protested that a complete change of the vehicle would be immensely costly, and that significant, and immediate, gains could be had by switching to hybrids with auto start and stop facilities to eliminate emissions at traffic lights and in traffic jams.

But Goyal dismissed hybrid vehicles as 'an intermediate technology, that the country has no use for'. India, he said, should opt for 'the best available technology'. Taking a dig at Maruti, he said, 'A few manufacturers are promoting hybrid vehicles because they do not have electric vehicles in their portfolio'.[37] To provide financial incentives for the shift into EVs, the central government developed a Faster Adoption and Manufacturing of Electric Vehicles (FAME) programme. This came into force on 1 April 2015 and was extended as FAME-II on 1 April 2019 to last till 31 March 2022. FAME-I concentrated on determining the incentives manufacturers would need to make the switch. As part of this, it mapped out a plan for establishing charging stations throughout the country, concentrating first on major cities and highways. Addressing this objective, it set up 427 charging stations on an experimental basis.

FAME-II began the transition out of liquid fuels in earnest. It allocated an overall budgetary support of $1.4 billion (Rs 100 billion) to encourage the purchase of 1 million electricity-fuelled motorcycles and scooters (two-wheelers), 500,000 electric rickshaws (three-wheelers), 35,000 E-automobiles, and 8,000 E-buses, all of this to be accomplished in its final year, by 31 March 2022.[38] In the first 2 years of FAME-II, only 5% of the $1.4 billion set aside for subsidies was disbursed and only 5,600 automobiles and 58,613 electric motorcycles and scooters were sold under the scheme.[39] So, in June 2021, the government increased the subsidies and extended the scheme for a further 2 years to 31 March 2024. This had an immediate impact. On 28 July, Tata Motors, which has captured 85% of the E-automobile market so far, reported that 650 of the 8,033 Nexons—its most popular mini-SUV—sold in June had E-engines. The company ascribed the breakthrough to the fact that the central and state government subsidies had brought the cost of the E-variant down to the point where it was only Rs 200,000–300,000 ($2,600–3,900) more expensive than the diesel-engined variant. Since the running cost of the E-Nexon was only a sixth of the diesel variant, even buyers who drove as little as 40 km a day could now recover the extra capital cost of the car in just over 2 years.

The additional subsidies, combined with a recovery from the COVID-19 lockdown, caused a sharp increase in EV sales. In 2021–22, E-automobile sales trebled from a little over 5,000 to 17,802; motorcycle and scooter sales went up five-fold to 231,000; and E-rickshaws doubled from 95,000 to 177,874. These achievements fell short of targets by a wide margin—the goal for two-wheelers, for instance, had been a million vehicles—but were nevertheless a good start.

But in the automobile market, the FAME-II targets had been 35,000 E-vehicles. Despite the very large subsidies available, the total sales in 3 years came to only half that number. The main reason was the consumers' sensitivity to price. In 2020, the showroom price of the basic E-Nexon had been Rs 1.43 million ($19,000) and that of the diesel variant Rs 8.3 lakh ($11,100). The price difference between the two was therefore $7,900. Tata was able to bring this down by two-thirds only because of the augmented subsidies given under the FAME-II programme in 2021 together with state government subsidies available from Delhi, Maharashtra, Gujarat, and Karnataka, and from Meghalaya in the north-east. Together, these subsidies added up to approximately 30% of the price of the car.

In 2021–22, compared with 17,100 EVs, the total number of cars sold was 3.09 million. Of these, only 2%—around 150,000—were vehicles costing more than Rs 1 million ($13,300). But none of the EVs now being offered in the market costs less than Rs 1.2 million ($16,000).

The main benefit of the FAME programme lies therefore not in its capacity to replace liquid-fuelled vehicles with electric alternatives, but in the early start it has given the automobile industry in domesticating an entirely new technology that might enable it to capture a large share of the world market for EVs. It is therefore a replica of the market strategy that China used to capture 80% of the world market for solar panels. This is reflected by a paragraph in the report on electrification that NITI Aayog unveiled at the annual meeting of the Society of Indian Automobile Manufacturers (SIAM) on 7 September 2018. India, it stated has, 'A relative abundance of exploitable renewable energy resources and *a high availability of skilled manpower and technology in manufacturing and IT software*' (emphasis added). This suggests that the government's real purpose was to get an early start in a sunrise industry, where India's low employee costs and established engineering skills in automobile (specially components) manufacture and digital technology would allow it to carve a place for itself in the *world* market for EV automobiles.[40]

In the home market, further rapid penetration will depend on the speed with which the cost of EVs can be brought down and their range increased. Both will doubtlessly happen, but it is unlikely that the price of EVs will come down to that of equivalent liquid-fuelled automobiles.

Three Indian companies launched EVs in December 2019 to January 2020 and twelve more have announced their intention to do so. But their ex-show-room prices range from Rs 1.5 million for a small car with a 21.5-kWh battery capacity and a 213-km range to Rs 2.5 million for a 39-kWh engine car with a claimed range of 452 km.[41] All of these are between two and three times the price of standard cars with similar dimensions and performance but much greater ranges. Because the battery makes up such a large part of their weight, EVs also weigh on average one-and-a-half times as much as their standard fuel equivalents. For instance, the Tata Nexon EV weighs 1,400 kg, but the Suzuki Swift, also made and sold in India and giving about the same performance with a standard engine

Table 8.3 Electric cars launched in India in 2020.

Car	Engine (KWh)	Range (km)	Price (Rs million and USD)
Tata Tigor	21.5	213	1.2 ($13,300)
MG ZS-EV	45	340	2.09 ($27,800)
Hyundai Kona	39	452	2.38 ($31,000)

Source: Prepared by authors from current data.

Table 8.4 Cost comparison between a standard Tata Nexon car and EV version.

Car	Engine	Range (km)	Price (Rs million and USD)
Tata Nexon XMA-AMT	1.2 litres 118 bhp	500 plus	($11,100)
Tata Nexon EV	30 kWh 127 bhp	312	$19,000)

Source: Prepared by authors from current data.

but with a much longer range, weighs 885 kg. At the other end of the scale, the Tesla S weighs 2.25 tonnes. The Mercedes 300 weighs 1.68 tonnes. Tables 8.3 and 8.4 show that whatever else electric cars may achieve for India, they will do next to nothing to reduce greenhouse gas emissions and reduce India's carbon footprint. Even the smallest of the EVs on offer in 2020 falls into the premium car range that accounts for only one in seven of all cars sold in the country and only 2% of the annual sales of all motorized vehicles.

Can these prices fall much further? The answer is almost certainly yes, in large part because of the benefits of mass production, but not by much. Everything in the EVs apart from the engine will be the same as in current liquid-fuelled cars, so these benefits have already been realized. The only new component is the engine. But here too, in key components like lithium the bulk of the reduction in cost has already taken place. Today it is hard to remember that in 2010 a lithium-ion car battery cost $1,183 per kWh, because by 2019 this had fallen to $156.[42] Most industry analysts expect the cost to fall further and stabilize eventually at around $100 per kWh by 2025. That will bring the cost of a standard 30-kWh power train down from $5,000 in 2019–20 to $3,500 by 2025. Promoters of EVs claim that at this cost EVs could become competitive with standard vehicles[43] because of their lower running costs. But that advantage will be more pronounced in capital-rich countries, where interest rates are low to negligible, as compared to developing countries such as India, where much of the lower running cost will have to be offset against the higher interest charged on car loans. Table 8.4 shows why this is unlikely to happen in India. The two Tata automobiles it depicts are identical except for the engine, and both the engines develop nearly the same horsepower. So, the entire $8,000 difference in price is accounted for by the much higher cost of the electric version's engine and the portion of the discounted net present value of the fuel cost that the company expects the buyer to save, which it feels it can appropriate. On the other hand, one food delivery service in India has committed to purchasing more than 160,000 EVs, a potential cause for optimism.

Process heat for industry

Surprising as it may appear, up to 2018, India's planners had given no thought to the need for process heat in industry. It was only in that year, when Europe and the US woke up to the need and turned to green hydrogen for a possible answer, that they hurriedly did the same. The potential for generating green hydrogen in India is discussed in chapters 10 and 12.

Notes and References

1. Solar Power in India. https://en.wikipedia.org/wiki/Solar_power_in_India#cite_note-189 (viewed 3/5/2022).
2. The official exchange rate on 31 January was Rs 46.226 to the dollar.
3. Ibid.
4. Number of hours of the year that the plants are actually generating electricity.
5. Solar Power in India, https://en.wikipedia.org/wiki/Solar_power_in_India#cite_note-91; and *The New Indian Express*, 6 May 2022, https://www.newin

dianexpress.com/business/2022/May/06/india-installed-10-gw-of-solar-capacity-in-2021-report-2450317.html.

6. Uma Gupta: India added a record 13.5 GW of solar and wind capacity in 2021–22, *pv magazine*, 25 April 2022. https://www.pv-magazine.com/2022/04/25/india-added-a-record-13-5-gw-of-solar-and-wind-capacity-in-2021-22/ (viewed 3/5/2022).
7. Uma Gupta: India may add 12.5 GW of solar in 2022–23, *pv magazine*, 10 January 2022. https://www.pv-magazine.com/2022/01/10/india-may-add-12-5gw-of-solar-in-2022-23/ (viewed 3/5/2022).
8. List of solar thermal power stations. Wikipedia, 21 October 2022. https://en.wikipedia.org/wiki/List_of_solar_thermal_power_stations (viewed 4/5/2022).
9. Solar Energy Technologies Office: Generation 3 Concentrating Solar Power Systems (Gen3 CSP). https://www.energy.gov/eere/solar/generation-3-concentrating-solar-power-systems-gen3-csp (viewed 4/5/2022).
10. Indian Wind Energy Association: Wind Power Potential. http://www.inwea.org/wind-energy-in-india/wind-power-potential/ (viewed 5/5/2022).
11. In the US, the average height of onshore turbines is 139 m (466 ft) and the tallest is 172 m (574 ft). https://www.vox.com/energy-and-environment/2018/3/8/17084158/wind-turbine-power-energy-blades (viewed 5/5/2022). Offshore turbines are even larger. GE is building a 12-MW wind turbine system that will be 266 m (873 ft) tall and have blades 107 m (351 ft) in length. David Roberts, VOX online, 20 May 2019.
12. Hossain, Jami et al. (2016). Re-assessment of wind energy potential with new technology in India. *Wind Engineering* 40(4):379–97. https://doi.org/10.1177/0309524X16651176 (viewed 6/5/2022).
13. The length of the turbine blade varies from 37 to 42 m for turbines generating 2 MW of power, so one blade diameter, after including the core, is roughly 90 m.
14. India Adds 275 MW of New Wind Capacity in Q1 2022; Gujarat Leads. https://mercomindia.com/india-adds-wind-capacity-q1-2022/#:~:text=India%20added%20275%20MW%20wind,of%20New%20and%20Renewable%20Energy (viewed 6/5/2022).
15. Uma Gupta: Gujarat 700 MW award cancellation will impact solar investment, *pv magazine*, 12 February 2021. https://www.pv-magazine-india.com/2021/02/12/gujarat-700-mw-award-cancellation-will-impact-solar-investment/ (viewed 6/5/2022).
16. Mayank Agarawal and Kundan Pandey: As States Back Out of Solar Contracts, India's Green Energy Targets at Jeopardy. https://thewire.in/energy/as-states-back-out-of-solar-contracts-indias-green-energy-targets-at-jeopardy (viewed 11/5/2022).
17. Rakesh Ranjan: Uttar Pradesh Scraps 500 MW Solar Auction, Plans to Re-Auction Hoping for Lower Bids, 3 June 2021. https://mercomindia.com/uttar-pradesh-scraps-500-mw-solar-auction-plans-to-re-auction-hoping-for-lower-bids/ (viewed 11/5/2022).

18. Ravleen Kaur: A decade on, India's first solar park has many promises left to fulfil, 18 March 2022. https://india.mongabay.com/2022/03/a-decade-on-indias-first-solar-park-has-many-promises-left-to-fulfil/ (viewed 12/5/2022).
19. Lydia Powell, Akhilesh Sati, and Vinod Kumar Tomar: Solar Energy in India: Contestation of Land Rights, 3 May 2022. https://www.orfonline.org/expert-speak/solar-energy-in-india/ (viewed 11/5/2022).
20. https://en.wikipedia.org/wiki/Renewable_energy_in_India: Year wise energy generation trend (accessed 9/1/2025).
21. Dilip Kumar Jha: India to achieve 7.2% of ethanol blending with petrol this season, *Business Standard*, 3 April 2019. https://www.business-standard.com/article/economy-policy/india-to-achieve-7-2-of-ethanol-blending-with-petrol-this-season-119040300816_1.html (viewed 12/5/2022).
22. *The Economic Times*, 3 April 2019. https://economictimes.indiatimes.com/industry/energy/oil-gas/ethanol-blending-with-petrol-may-reach-7-2-pc-in-2018-19-season-from-4-2-last-year/articleshow/68705372.cms?from=mdr (viewed 13/5/2022).
23. Indian Petroleum and Natural Gas Statistics 2018–19, p. 91. https://www.data.gov.in/catalog/indian-petroleum-and-natural-gas-statistics-2018-19 (viewed 13/5/2022).
24. Reuters, 11 February 2020. https://www.livemint.com/market/commodities/india-sugar-output-for-2019-20-unlikely-to-see-big-revision-isma-11581415873956.html (viewed 13/5/2022). India sugar output for 2019–20 unlikely to see big revision: ISMA.
25. http://petroleum.nic.in/sites/default/files/biofuelpolicy2018_1.pdf (viewed 17/5/2022). The precise listing of usable biomass is as follows: 'Potential domestic raw materials for production of biofuels in the Country are, for Ethanol Production: B-Molasses, Sugarcane juice, biomass in form of grasses, agriculture residues (Rice straw, cotton stalk, corn cobs, saw dust, bagasse etc.), sugar containing materials like sugar beet, sweet sorghum, etc. and starch containing materials such as corn, cassava, rotten potatoes etc., Damaged food grains like wheat, broken rice etc. which are unfit for human consumption, Food grains during surplus phase.'
26. Department of Agriculture and Food, Western Australia (2006): Ethanol production from grain. https://researchlibrary.agric.wa.gov.au/cgi/viewcontent.cgi?article=1031&context=pubns (viewed 18/5/2022).
27. What is still limiting the deployment of cellulosic ethanol? Analysis of the current status of the sector. https://publications.jrc.ec.europa.eu/repository/handle/JRC1179555
28. For a detailed description of how corn fibre is used to produce ethanol see Kurambhatti, Chinmay V., et al. (2018). Ethanol production from corn fiber separated after liquefaction in the dry grind process. Energies 11(11):2921. https://www.mdpi.com/1996-1073/11/11/2921 (viewed 18/5/2022).

29. Annual report 2018–19. https://praj.net/wp-content/uploads/2019/06/praj-annual-report-2018-19.pdf (viewed 19/5/2022). This site has closed. The current website is https://www.google.com/search?q=praj+industries+ethanol&sca_esv=0d3dcdd58ef81c8a&sxsrf=ADLYWIKc3QdgYssNuxVNN4uHtoKh4igpdg%3A1736372817057&ei=UfJ-Z6aYA-DKseMPweLOwQE&oq=+Praj+industries+&gs_lp=Egxnd3Mtd2l6LXNlcnAiESBQcmFqIGluZHVzdHJpZXMgKgIIATIOEAAYgAQYkQIYyQMYigUyCxAAGIAEGJECGIoFMgsQABiABBixAxiDATIFEAAYgAQyCxAAGIAEGLEDGIMBMgUQABiABDILEAAYgAQYkgMYigUyCxAAGIAEGJIDGIoFMgUQABiABDIIEAAYgAQYsQNI3npQtThYtVhwAXgBkAEAmAHuAaABlgSqAQUwLjIuMbgBAcgBAPgBAZgCBKACrQTCAgoQABiwAxjWBBhHwgIIEAAYgAQYogTCAgYQABgWGB7CAgsQABiABBiGAxiKBZgDAIgGAZAGCJIHBTEuMi4xoAeAEg&sclient=gws-wiz-serp (accessed 09/01/2025).
30. Booklet of the 2nd EU–India Conference on Advanced Biofuels, New Delhi, 11–13 March 2019. http://artfuelsforum.eu/wp-content/uploads/2019/04/Booklet-on-technologies-presented-at-the-2nd-EU-India-Conference-on-Advanced-Biofuels.pdf (viewed 20/5/2022).
31. Prashant Krar: Paddy stubble to drive bio-fuel production in Punjab, Haryana, *The Economic Times*, 12 November 2019. https://economictimes.indiatimes.com/industry/energy/oil-gas/paddy-stubble-to-drive-bio-fuel-production-in-punjab-haryana/articleshow/71984431.cms?from=mdr (viewed 19/5/2022).
32. India Glycols launches second-generation ethanol plant in Kashipur, *Biofuels International Magazine*, 2016. https://biofuels-news.com/news/india-glycols-launches-second-generation-ethanol-plant-in-kashipur/ (viewed 20/5/2022).
33. Prashant Krar: Paddy stubble to drive bio-fuel production in Punjab, Haryana, *The Economic Times*, 12 November 2019. The para is talking about ethanol plants being set up using the fermentation route in Punjab and Haryana. And the footnote opened easily. https://economictimes.indiatimes.com/industry/energy/oil-gas/paddy-stubble-to-drive-bio-fuel-production-in-punjab-haryana/articleshow/71984431.cms?from=mdr (viewed 19/5/2022).
34. Yuanrong Zhou, Stephanie Searle, and Sunitha Anup: Techno-economic analysis of cellulosic ethanol in India using agricultural residues. https://theicct.org/sites/default/files/publications/cellulosic-ethanol-analysis-india-jun2021.pdf (viewed 20/5/2022).
35. India Leaps Ahead: Transformative Mobility Solutions for All. Report published by NITI Aayog: Government of India. New Delhi. http://niti.gov.in/writereaddata/files/document_publication/RMI_India_Report_web.pdf (viewed 23/5/2022).
36. *The Hindu*, 18 August 2017: India to sell only electric vehicles by 2030: statement by transport minister Piyush Goyal., https://www.thehindu.com/news/national/karnataka/india-to-sell-only-electric-vehicles-by-2030-piyush-goyal/article19516175.ece. Accessed (viewed 20/5/2022).

37. Ibid. https://www.business-standard.com/article/automobile/auto-sales-remain-low-in-fy22-two-wheeler-dispatches-lowest-in-10-years-122041300915_1.html
38. *Business Standard*, 13 April 2022; and https://auto.hindustantimes.com/auto/news/electric-vehicle-retail-sales-in-india-witness-three-fold-jump-in-fy22-fada-41649571975193.html#:~:text=Total%20electric%20vehicle%20(EV)%20retails,the%20financial%20year%202020%2D21, HT Auto, 10 April 2022 (viewed 23/5/2022).
39. FAME-2 incentive scheme for electric vehicles extended by two more years, *The New Indian Express*, 26 June 2021. https://www.newindianexpress.com/business/2021/Jun/26/fame-2-incentive-scheme-for-electric-vehicles-extended-by-two-more-years-2321671.html
40. Are all quotes from the NITI report. https://niti.gov.in/writereaddata/files/document_publication/EV_report.pdf (viewed 24/5/2022).
41. Details of these are now available on dealers' websites such as https://www.carwale.com/mg-ca399rs/zs-ev/price-in-delhi/?qti=1(viewed 24/5/2022).
42. Paul Caine: Falling Battery Price Transforms Economics of Green Energy. *WTTW Science News*, 2 January 2020. https://www.newindianexpress.com/business/2021/Jun/26/fame-2-incentive-scheme-for-electric-vehicles-extended-by-two-more-years-2321671.html (accessed 09/01/2025).
43. Ibid. George Crabtree, senior scientist and distinguished fellow at Argonne National Laboratory in Lemont, Illinois: in a video interview that is part of the above article.

9
India's Wind and Solar Power Potential

Wind

This chapter is intended to define the physical potential for production of power from a combination of wind and solar sources of energy in India. In the case of wind, we focus on the electricity generation potential of onshore winds and base our estimates of power generation upon the yield from a fleet of 2.5-MW turbines manufactured by GE (General Electric) the most commonly deployed turbines in the world today. In India, except in a few locations along the Coast of Gujarat, the potential for generating power from offshore wind is small because of the relatively steep descent of the continental shelf into deeper waters.

With very few exceptions, therefore, Wind turbines are almost all located onshore, and begin to generate power at wind speeds greater than 3.0 metres per second. We feel such a wind velocity as a light breeze, blowing at just under 11 km/h. For safety reasons the turbines are also designed to cut out when the speed exceeds 25 m/s. or 91 kms/hour. These turbines are mounted at a hub height of 110 m above the ground and the blades (three of them) are each 56.9 m long. The lowest point on their trajectory is thus over 40 m above the ground. The generation of power of the turbine as a function of wind speed is defined by what is known as the power curve. This is illustrated in Figure 9.1.

Solar

Our estimate of India's Solar Photo-Voltaic (SPV) power potential is based on the use of fixed-tilt polysilicon PV modules which convert 16.2% of the incident solar radiation into electricity. For estimating both

Sun, Wind, and Biomass. Prem Shankar Jha and Michael B. McElroy, Oxford University Press.
 DOI: 10.1093/9780198944768.003.0009

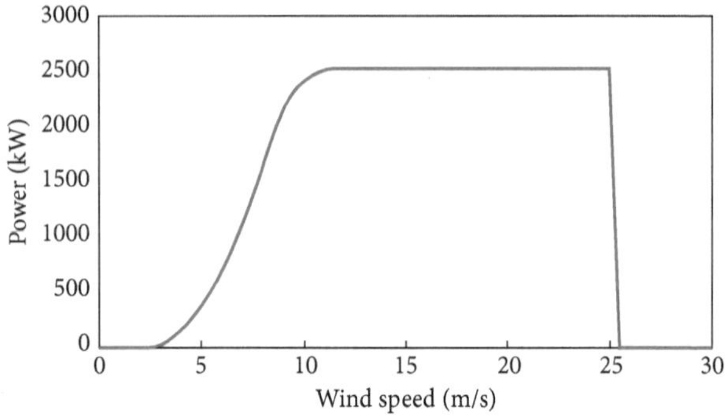

Figure 9.1 Wind power curve for GE 2.5-MW turbine.
Source: Wikimedia Commons.

wind and solar power generation capacities we confined our attention to the generation of power from large, utility-scale, facilities designed to feed electricity into the national grid. We did not explicitly consider Rooftop solar generation capacity, but recognise that it can make an important contribution in the future if the incentives given to it are generous and their free access to sunshine is guaranteed by limiting the heights of adjacent buildings, to prevent them from cutting off the sunshine in the future. The results obtained and the sources we have relied on to obtain them are described in greater detail below.

Estimating the power obtainable from wind

Wind speeds are variable in both space and time. The wind speed data we employed in this study to estimate the power generation potential of wind were obtained from the US—MERRA-2 (Modern-Era Retrospective Analysis 'for Research and Applications, version 2). This is an estimation process available from NASA's Goddard Earth Sciences Data and Information Services Center.[1] MERRA-2 is an ex-post reconstruction of the state of the future atmospheric condition that is based upon the reanalysis, of past weather patterns, MERRA-2 employs essentially the same techniques used by meteorologists to develop weather forecasts, but with, the advantage that the forecasts can be updated regularly by comparing them to what actually happened when similar conditions had developed in the past, using real moment-by-moment observational data to confirm or modulate the current forecast.

The record of former weather patterns is collected individually for grids with a horizontal resolution of 0.5 degrees latitude and 0.67 degrees of longitude. For our present purpose, which is to predict the electricity likely to be available from solar and wind power in the near future, we have compared our hourly observations with the actual data on sun and wind availability available hour by hour in 2019. Wind speeds intercepted by the rotors of the turbines were evaluated by extrapolating the MERRA data which is for 10 m to 50 m levels above ground, to estimate wind speeds at 40 metres to 167 metres above it, for these are the levels that correspond to the upper and lower limits of the airspace occupied by the blades of the 2.5 MW turbines. Lu et al. made their analysis based on MERRA data for 2016. The choice of a particular year, 2016 or 2019, is not expected, however, to markedly impact the results obtained from the present analysis. The selection of 2019 for the current study was motivated by the choice of a baseline for the analysis of decarbonisation that is more fully described in chapter 10.

Consistent with previous studies of the opportunity that exists for harnessing wind and solar energy in India, we chose to exclude areas that are forested, urban, or covered with either water or ice, in addition to regions identified with slopes of more than 20% and/or elevations greater than 3,000 m. We assumed further that only 30% of cropland and 80% of rangeland could practically be assigned to the development of wind farms. The distribution of land area types was identified based on data from NASA's Moderate Resolution Imaging Spectroradiometer (MODIS) satellite MCD12C1,[2] as reconstructed by Lu et al., with results as displayed in Figure 9.2. Slope data were derived from the NASA Shuttle Radar Topography Mission (SRTM).[3] Turbines were separated by nine rotor diameters to minimize turbine–turbine interactions. This implies placement of at most one turbine for every 0.64 km^2 of land judged suitable for wind power development.

The distribution of wind power generation capacity factors derived from the above analysis is illustrated on a national basis in Figure 9.3a. Capacity factors refer to the fraction of the power produced by a turbine relative to its theoretical maximum. For example, if the wind speed for every hour of the year were to lie on the flat part of the power curve

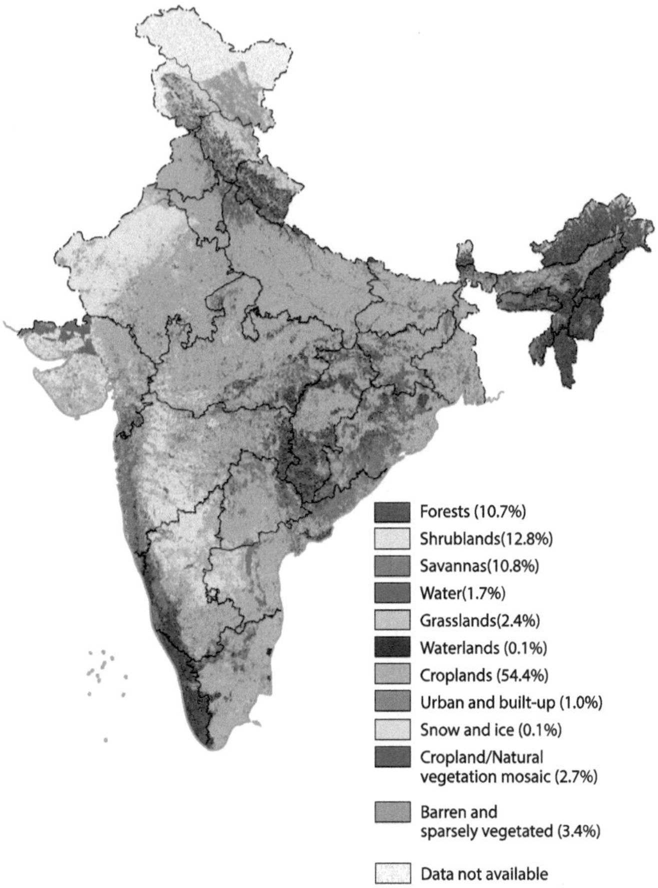

Figure 9.2 Land type classifications determined by MODIS satellite MCD12C1.

Source: Recreated by the authors based on dataset from Lu et al.; Tianguang, Lu., et al. India's potential for integrating solar and on- and offshore wind power into its energy system. *Nature Communications* 11, 1–10 (2020).

presented in Figure 9.1, the resulting capacity factor would be equal to 100%. Capacity factors—averages—are presented on a state-by-state basis in Figure 9.3b. They range from a high of 39.3% in Gujarat to a low of less than 4% in Ladaka and Arunachal Pradesh. In general, capacity factors and potential power sources for wind are highest in the west and south, and lowest in the north and east.

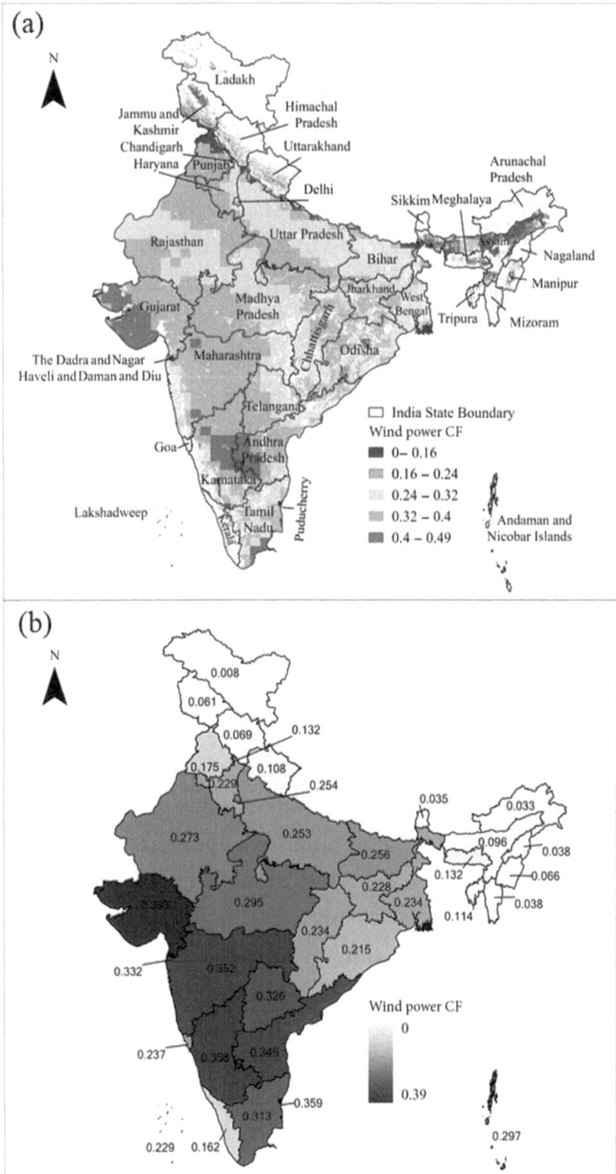

Figure 9.3 National spatial distribution (a) and state-by-state distribution (b) of mean capacity factors (CUF) for onshore wind, 2019.

Note: Not to scale; for representational purposes only.

We have presented State-by-state estimates of potential generating capacities in Figure 9.4a, with results for corresponding annual power production in terawatt hours (TWh) displayed in Figure 9.4b. Our estimates show that potential installed capacities range from a high of 817 GW in Rajasthan, 499 GW and 460 GW in the neighbouring states of Madhya Pradesh and Maharashtra respectively. Not surprisingly, therefore, Rajasthan, Maharashtra, and Madhya Pradesh, rate highest in terms of their capacity to generate potential power production, with estimated annual yields of 1,958, 1,420, and 1,288 TWh respectively. The potential national annual total could be as high as 10,320 TWh, which is almost double the estimate made by NITI Aayog[4] of a potential demand of 5,651 TWh in 2047 This is substantially larger than the estimate made by The Energy and Resources Institute (TERI) and Shell[5] of 8,100 TWh in 2051, and by Song et al.[6] of 8,323 TWh in 2050. Given the results displayed in Figures 9.3 and 9.4, it would be tempting to conclude that onshore wind in India should be more than sufficient to meet the projected future countrywide demand for electricity. The comparison, however, would be misleading.

The supply of electricity must be tailored to meet demand in effectively real time. The variation in the source from wind is unlikely to match up as closely as necessary with the accompanying pattern of demand. There will be occasions when the supply from wind may be significantly greater than demand, not only in time but also in space, offset by times when it may be seriously deficient. The overall power system will need to deal with the input from wind and from the solar source discussed below and also accommodate contributions from a range of existing background ancillary sources, including, for example, thermal, hydro, and potentially nuclear. And it must do so in the face of a complex pattern of demand in both space and time.

We take up this issue in more detail in Chapter 11 which outlines a strategy based on the analysis by Song et al. for decarbonization of the Indian economy by 2050. An important part of this strategy involves use of wind and solar power when available in excess to provide a source of what is referred to as green hydrogen, hydrogen formed by using electricity to split the water molecule into its component hydrogen and oxygen molecules, a process referred to as electrolysis. Storage and release of electrical energy in the form of hydrogen can be used to balance overall

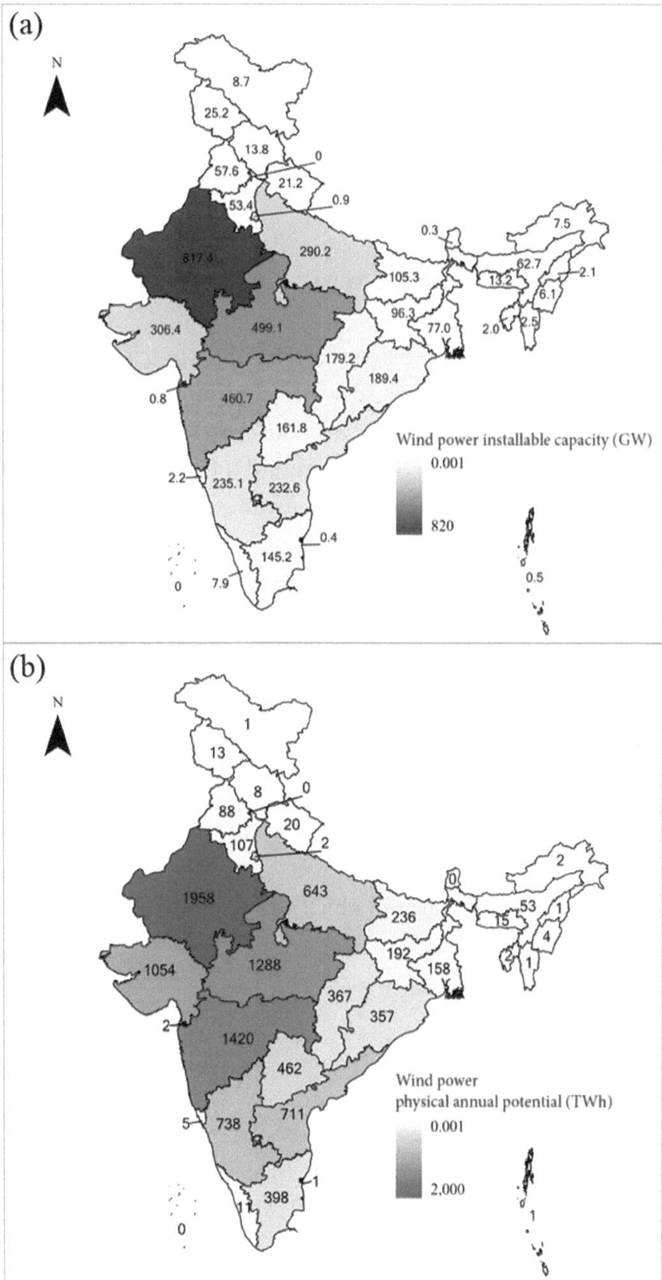

Figure 9.4 State-by-state distribution of potential installed capacities (a) and annual power production (b) for onshore wind, 2019.

Note: Not to scale; for representational purposes only.

demands for electricity both regionally and nationally. Hydrogen can be employed further, as discussed in Chapter 10, to reduce the demands for fossil energy in portions of the transportation sector—heavy trucks and buses for example—and in selected components of industry, in the so-called hard-to-abate sectors such as iron and steel and cement.

Estimating the usable energy obtained from the sun

The placement of an array of solar panels requires a much greater commitment of land than positioning a set of wind turbines capable of yielding a comparable source of electric power: the spatial footprint of the wind turbine is small compared to the footprint of the comparable solar array. In assigning areas for potential development of utility-scale solar power we elected to exclude forested regions and regions covered with either water or ice, as we did with the selection of areas for the siting of wind resources. But, in addition, we eliminated croplands and regions identified with a combination of cropland and natural vegetation. Suitability factors ranging from 5% to 20% were assigned to different types of the remaining land, as indicated in Table 9.1. The minimum solar radiation required for the development of a solar installation was set at 1,400 kWh per square metre per year. Regions with slopes greater than 5% were excluded.

Table 9.1 The table below, which sets out the present land used in India.

Land use type	Setting (%)
All forest	0
Closed shrublands, woody savannas	5
Grasslands, open shrublands, savannas	20
Barren or sparsely vegetated	20
Croplands, cropland/natural vegetation mosaic	0
Urban and built-up	5
Waterbodies, permanent wetlands, snow and ice	0

Source: From Tianguang, Lu, et al. (2020). India's potential for integrating solar and on- and offshore wind power into its energy system. *Nature Communications* 11:1–10.

The yield of power from a particular solar installation depends not only on the availability of sunlight over the course of the year but also on the packing density of the solar panels. This is defined by the distance between individual arrays of panels. To maximize their output, it is important to minimize the shading contributed by neighbouring panels. India is located north of the equator. So the panels should face to the south, casting shadows therefore to the north. To determine the optimal distance between adjacent rows of panels, we considered the elevation of the sun at 3 pm on 21 December (the winter solstice). That is when the shadows are longest and the shading greatest. Capacity factors for individual solar arrays are defined as the ratio of the electricity generated on an annual basis relative to the maximum physical potential.

Results for solar irradiation employed in this analysis were taken from the NASA GEOS-5 FP database[7] which reports hourly temperatures and incident solar radiation with a spatial resolution of 0.25 degrees latitude by 0.31 degrees longitude. Defining the diurnal, and more importantly the seasonal, trends in temperature is important for estimating the output of SPV panels, reflecting the fact that their efficiency declines at higher temperature. Taking all these factors into account, our analysis suggests a generation potential for solar power from a utility-scale PV of 22,170 TWh. This is more than twice the 10,320 TWh inferred for onshore wind.

Results for the annual average range of potential capacity factors for utility-scale SPV judged socially and economically acceptable based on the criteria outlined above are presented in Figure 9.5a. A breakdown of the output on a state-by-state basis is presented in Figure 9.5b, with data for related average potential capacities (GW) and annual power yields (TWh) displayed in Figures 9.6a and 9.6b respectively. The contributions from wind and sun are not, however, distributed uniformly in either time or space. The source from wind peaks typically during the monsoon season, while the contributions from solar are more important during spring and autumn, a pattern discussed further in the decarbonization chapter that follows (Chapter 10). A notable feature of the results is that the state with the highest potential for generating power from utility-scale SPV is the same as the state that ranks highest for wind, Rajasthan. Its potential of 7,120 TWh is far greater than the potential inferred earlier for wind, 1,958 TWh.

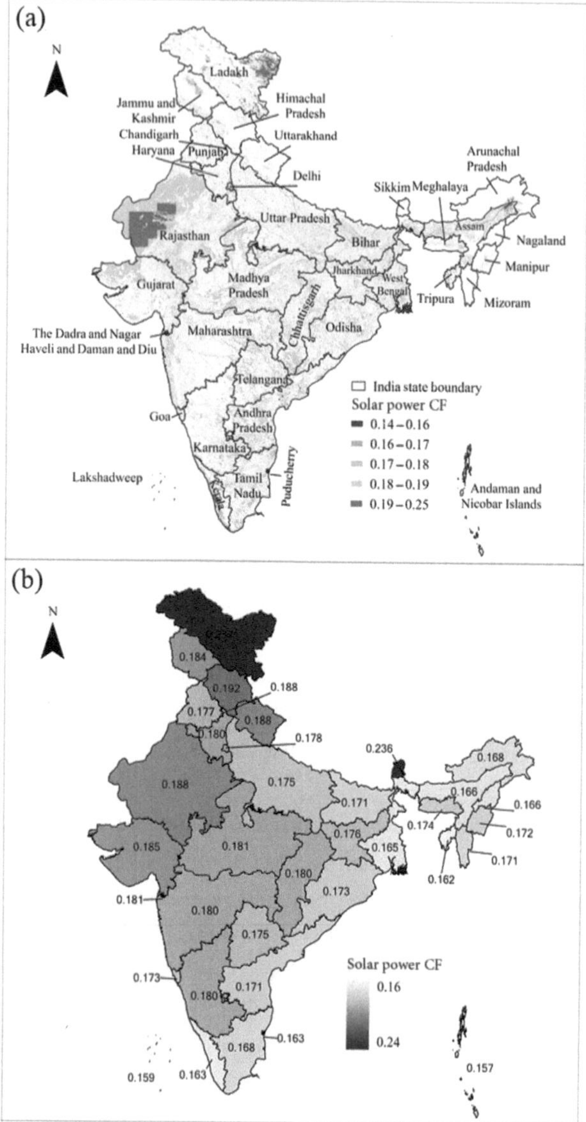

Figure 9.5 National spatial distribution (a) and state-by-state distribution (b) of mean capacity factors for SPV, 2019.

Note: Not to scale; for representational purposes only.

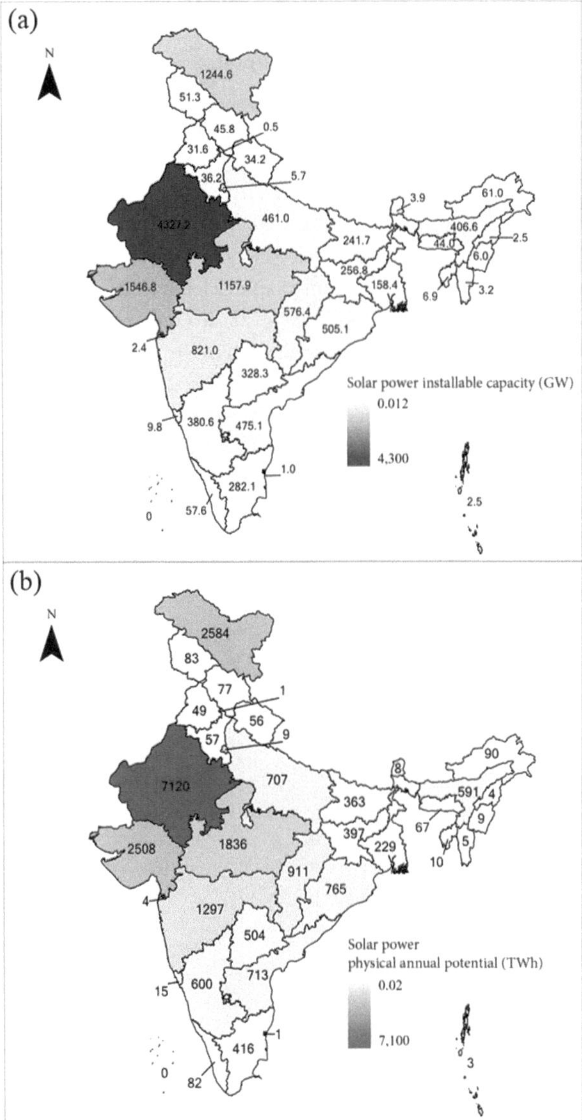

Figure 9.6 State-by-state distribution of potential installed capacities (a) and annual production of power (b) for SPV, 2019.

Note: Not to scale, for representational purposes only.

Notes and References

1. Gelaroa, R., et al. (2017). The Modern-Era Retrospective Analysis for Research and Applications, version 2 (MERRA-2). *Journal of Climate* 30:5419–554.
2. Jarvis, A., et al. (2008). Hole-filled seamless SRTM data V4. http://srtm.csi.cgiar.org (viewed 9/6/2022).
3. NITI Aayog: Energizing India, 2017. https://niti.gov.in/sites/default/files/energy/Energising-India.pdf (viewed 14/6/2022).
4. TERI/Shell: India: Transforming to a Net-Zero Emissions Energy System, 2021. chrome-extension://efaidnbmnnnibpcajpcglclefindmkaj/https://www.niti.gov.in/sites/default/files/2022-12/NEP-ID_27.06.2017.pdf.pdf (viewed 14/6/2022).
5. Song, S., et al. (2022). Deep decarbonization of the Indian economy: 2050 prospects for wind, solar, and green hydrogen. *iScience* 25(6).
6. Global Modeling and Assimilation Office. File Specification for GEOS-5 FP (Forward processing), 2017. https://gmao.gsfc.nasa.gov/products/documents/GEOS_5_FP_File_SpecificationON4v1_1.pdf (viewed 22/6/2022).
7. https://www.nccs.nasa.gov/services/data-collections/coupled-products/geos5-forecast (accessed 13/01/25).

10

Potential for Green Hydrogen in India: Green Versus Grey Hydrogen

Classification of hydrogen as green refers to the manner of its production. To merit the label green, the hydrogen must be produced without a concurrent source of CO_2. Electrolysis, running a current of electricity through water and splitting the water molecule into a combination of H_2 and O_2, provides the most common contemporary source of green hydrogen. The bulk of the hydrogen on the current world market, however, is not green. Most of it is produced primarily from natural gas, with some production, particularly in China, from coal. The hydrogen formed by cracking natural gas is associated with production of about 10 kg of CO_2 for every kilogramme product of H_2. It is described as grey if the product CO_2 is released to the atmosphere and blue if it is captured and sequestered. Most of the hydrogen produced and consumed globally today is grey. If it were produced from coal, rather than natural gas, the yield of CO_2 would be even greater.

In India, hydrogen is used mainly in the production of nitrogen fertilizer and in the oil refinery sector. The demand for these two sectors in 2019 amounted to 4.4 Mt. It is projected to grow, however, to as much as 136.7 Mt by 2050 under the conditions envisaged in the decarbonization study reported by Song et al.,[1] as discussed in Chapter 11. The Song analysis envisages not only substitution of green for grey hydrogen in the current industrial sector but also applications in a variety of other contexts, including deployment to compensate for the variability of renewable power sources such as wind and solar, as complement to battery-powered propulsion in portions of the transportation sector specifically for long-distance trucks and buses, deployment as a substitute for coking coal in the iron and steel industry, use as a source of high-temperature heat in the

Sun, Wind, and Biomass. Prem Shankar Jha and Michael B. McElroy, Oxford University Press.
 DOI: 10.1093/9780198944768.003.0010

cement industry, and potentially a variety of other applications in which it can serve as a substitute for carbon-emitting fossil energy.

Production of hydrogen: differences in costs and types of electrolysers

According to the IEA[2], the cost for grey hydrogen in India is currently about $1.8 per kg. If green hydrogen is to be competitive with grey in the years to come, it will be necessary for the cost to come down significantly, from the current level of $4.65 per kg as estimated by the International Renewable Energy Agency (IRENA)[3] to a level closer to the contemporary cost of the grey alternative. To meet this objective, IRENA suggests that the capital costs for electrolysers would need to fall by 80% from the present average of $650 per kWh to a level closer to $130. Their efficiency would need to rise from 65% to 76%; the duty cycle would need to increase by 30% from the current level of 3,200 hours per year to 4,200 hours per year; and their lifetime would need to double from the present 10 years to about 20 years. These are ambitious targets. However, the cost of providing and maintaining electrolysers represents only about a quarter of the cost of producing hydrogen by this route. Three-quarters of the cost is accounted for by the electricity required to break the hydrogen–oxygen bond in water.

An electrolyser is composed typically of an anode and a cathode separated by an electrolyte. There are three types of commercial and one still experimental electrolysers on the market at present. The simplest and the cheapest design uses an electrolyte consisting of an aqueous solution of potassium or sodium hydroxide as the medium through which to transfer the power required to split the water molecule. This is identified in what follows as the alkaline electrolysis cell (AEC). It operates at an efficiency of about 70%, and consumes therefore approximately 47 kWh of electricity to produce 1 kg of hydrogen (47 kWh/kg). By way of comparison, the energy content of hydrogen is 33.3 kWh/kg. A second design employs a solid specialty plastic as the electrolyte with a polymer membrane (PEM) operating currently at an efficiency of about 60%, requiring thus consumption of approximately 55 kWh of electricity to produce 1 kg of hydrogen (55 kWh/kg). Although more expensive today, the PEM electrolyser could prove the most successful in the long run, anticipating that its efficiency could be improved to 80% or even higher. Finally, there is a high-temperature electrolyser that uses ceramic membranes and superheated steam to split the water molecule. This would reduce the power required to as low as 41.4 kWh/kg of hydrogen, but without taking into account the energy that will be needed to superheat the water to the required temperature. In the foreseeable future, therefore, the market is likely to remain dominated by the AEC design, the cheapest and most widely used of the currently available options.

Momentum grows in the world for a green hydrogen future

The potential of 'green' hydrogen to replace 'grey' hydrogen in industrial processes that require large amounts of high-temperature process heat is being rapidly recognized all over the industrialized world. At the World Economic Forum in Davos in 2017, leaders of 13 global companies announced the formation of a Hydrogen Council. Three years later the Council had more than 40 members. In February 2018, the China Energy Corporation, joined by 18 other corporate sponsors, announced the formation of a China Hydrogen Alliance, with the goal of speeding up the development of China's hydrogen energy sector and rapidly improving its quality. In June 2018, France became the first country to formally announce its intention to produce green hydrogen. The following month, energy ministers from around the world gathered in Malmo, Finland and announced an initiative to develop green hydrogen and fuel cells as an alternative to CO_2-emitting gasoline and diesel in the transportation sector. Energy ministers of EU countries met in Linz, Austria in September 2018 and signed a declaration stating that green hydrogen would be a part of the EU's energy future. By the end of 2020, the EU and China had been joined by no fewer than 20 countries that announced similar commitments. In the same year, targets for the development of hydrogen in the US were outlined in the Hydrogen Program Plan published by the Department of Energy.[4] The goal, as stated in that plan, was to 'advance the production, transport, storage, and use of hydrogen across different sectors of the economy'.

The pace of progress in other elements of the global green hydrogen economy was highlighted by the opening by the Norwegian company Nel ASA in April 2022 of a fully automated, cost-cutting, 500-MW AEC electrolyser production facility in Heroya, Norway. According to the announcement, the Norwegian plant could be scaled up to 2 GW of annual production and could be expanded to 10 GW by 2025 'if required by the market'. In announcing the opening, commenting on the declining price of electrolysers, Nel's CEO Jon Andre Lokke stated that 'half of the savings will come from scale up and increased efficiency in production. The rest will come from economies of scale, and from effective industrial partnerships'. The goal of the company is to bring the levelized cost for green hydrogen down to \$1.50 per kg, assuming an electricity cost of \$20 per MWh. In the World Energy Outlook 2020, IEA[5] indicates that in the best locations with access to favourable policy support and finance, solar can now generate electricity 'at or below' \$20 per MWh. The IEA goes on to report that new utility-scale solar projects now cost \$30–60 per MWh in Europe and in the US, but costs are significantly lower, \$20–40 per MWh, in China and India, where 'revenue support mechanisms' such as guaranteed prices are in place. There are good reasons to believe that prices will continue to decline.

India joins the bandwagon

India joined the bandwagon in February 2021, with an announcement by its Finance Minister in her speech rolling out the budget that the government would set up a National Hydrogen Mission similar to the National Solar Mission that was created in 2009. In the budget she allocated Rs 8 billion (\$107 million) as seed money designed to leverage \$1.4 billion from the private sector to finance various ways of getting blue and green hydrogen and blending it with compressed natural gas to provide heat and to run the bus transport systems of Delhi and other major Indian cities. This inspired an Indian company, BGR Energy Systems, to team up with a European company, Fusion Fuel Green, to announce plans on 9 March 2021 to set up a demonstration plant at Cuddalore in Tamil Nadu.

In April 2021, the government organized an online Hydrogen Round Table with 25 panellists from 15 countries to discuss the ambitions it had for a fossil-free energy future. This was presided over by India's petroleum minister. The next step was to announce a draft policy on 6 August 2021 that promised to meet 10% of overall hydrogen needs of oil refineries by 2023–24 using green hydrogen. This commitment was to be raised to as much as 25% by 2028–29. Further, green hydrogen would be required to replace 5% of the grey alternative in the fertilizer industry and to supplant 5–20% of grey hydrogen in other, unspecified, industries. Nine days later, on 15 August on the 75th anniversary of India's independence, Prime Minister Modi caused a minor sensation when he announced the formation of a National Hydrogen Mission, committing the nation to making a 'quantum leap' in green hydrogen production that would enable it to 'capture the market for it as China had captured the market for solar photovoltaic panels'.

Green hydrogen from biomass

We choose in this chapter to focus on the production of green hydrogen using electricity sourced from carbon-free renewable sources exemplified by wind and solar. There is, however, an alternative means to produce this green hydrogen using biomass as a feedstock.

Plants absorb carbon dioxide from the atmosphere and use sunlight to convert it into their various chemical components, through a process known as photosynthesis. Gasification, which is controlled burning with a limited and carefully calibrated supply of either air or pure oxygen, yields 'synthesis gas', a mixture of carbon monoxide, hydrogen, and carbon dioxide along with small fractions of heavier gases such as methane. Today, this synthesis gas is the starting point for most of the petrochemical industries of the world. Gasification has been in use, with coal as the feedstock, for over to a century. It was first used to produce a fuel gas in 1812 by the London Light Gas and Coke company, mainly for street lighting. Later its uses expanded to providing heat and power, and finally by Germany during the second World War, to providing liquid fuel to their vehicles and aircraft.

In the past half century gasification been refined further, and made more versatile till synthesis gas can be obtained from virtually any form of biomass. The synthesis gas and oil it produces can replace coal and oil in most of their present downstream uses, but one of its more significant future uses could be the production of green hydrogen. What is important to note is that even the limited amount of CO_2 it generates reflects simply a recycling of some of the CO_2 that was originally present in the atmosphere. It therefore involves no net addition of CO_2. The potential for production of green hydrogen and other useful energy products from biomass is discussed further in Chapter 12.

Notes and References

1. Song, S., et al. (2022). Deep decarbonization of the Indian economy: 2050 prospects for wind, solar, and green hydrogen. *iScience* 25(6).
2. IEA: The Future of Hydrogen, 2019. https://www.iea.org/reports/the-future-of-hydrogen (viewed 12/7/2022).
3. IRENA: Green Hydrogen Cost Reduction: Scaling up Electrolysers to Meet the 1.5°C Climate Goal, 2020. https://www.irena.org/publications/2020/Dec/Green-hydrogen-cost-reduction (viewed 13/7/2022).
4. Menezes, M.W., et al. (2020). The US Department of Energy Hydrogen Program Plan. US Department of Energy.
5. IEA: World Energy Outlook 2020. https://www.iea.org/reports/world-energy-outlook-2020 (viewed 13/7/2022).

11

How Far Can India Go in Replacing Fossil Fuels by 2050?

As of 2022 India has become the world's third largest national emitter of CO_2, accounting for 7% of total global greenhouse gas emissions. In 2021 over 80% of India's energy needs were being met by three fuels: coal, oil and solid biomass. Coal has underpinned the expansion of electricity generation and industry, and remains the largest single fuel in the energy mix. Oil consumption and imports have grown rapidly on account of rising vehicle ownership and road transport use. Biomass, primarily fuelwood, makes up a declining share of the energy mix, but is still widely used as a cooking fuel. Despite recent success in expanding coverage of LPG in rural areas, 660 million Indians have not fully switched to modern, clean cooking fuels or technologies.[1]

Coal is responsible for 45% of the country's contemporary energy consumption, followed by petroleum and natural gas, 32%, with traditional biomass and organic waste accounting for an additional 20%. Renewable sources, including wind, solar, and hydro, account for only 2% of total current primary energy consumption. If India is to successfully navigate the transition to a low carbon future, it will be essential to cut back massively on its current reliance on coal, oil, and natural gas. As indicated in Chapter 9, wind and solar resources can play an important role in achieving this objective. So also can green hydrogen and biomass. The potential role of green hydrogen was discussed in Chapter 10. The immense contribution that biomass can make towards replacing fossil fuels is discussed in the chapters that follow.

Sun, Wind, and Biomass. Prem Shankar Jha and Michael B. McElroy, Oxford University Press.
 DOI: 10.1093/9780198944768.003.0011

Projections of India's future energy requirement

Over 80% of India's energy needs are met by three fuels: coal, oil and solid biomass. Coal has underpinned the expansion of electricity generation and industry, and remains the largest single fuel in the energy mix. Oil consumption and imports have grown rapidly on account of rising vehicle ownership and road transport use. Biomass, primarily fuelwood, makes up a declining share of the energy mix, but is still widely used as a cooking fuel. Despite recent success in expanding coverage of LPG in rural areas, 660 million Indians have not fully switched to modern, clean cooking fuels or technologies. Trends in GDP, population, and urbanization are the key drivers determining future demands for energy in the Indian economy.

In 2017, NITI Aayog, the Indian government entity responsible for drafting recommendations for policy, projected that India could maintain an annual growth of GDP as high as 8.5% out to 2042, decreasing subsequently to 7.4% over the period 2042–2047.[2] But over 30 years, from 1992 to 2021, through changes in policy, changes in government, economic booms, economic recessions, and the COVID-19 pandemic, the economy grew at an average annual rate of close to 6.25%.[3] For present purposes, following the analysis in Song et al.[4], we assume growth rates of 6% for 2022–2030, 6.5% for 2030–2040, and 5.7% for 2040–2050. These rates are consistent with recommendations by the IEA[5] and the United Nations Development Programme (UNDP).[6]

The population of India finally surpassed the population of China in 2023.The UNDP suggests that the Indian population will rise from 1.37 billion in 2019 to 1.6 billion in 2040, conclusions that are consistent with estimates by NITI Aayog[7] projecting an increase in the population of India to 1.68 billion by 2047. Assumptions for purposes of the present analysis with respect to the growth in GDP and population are summarized in Table 11.1.

Energy for transport

Demand for energy in the transport sector can be broken down in terms of requirements for passengers and freight. The controlling factors for

Table 11.1 Key socioeconomic assumptions for India.

	Unit	2020	2030	2040	2050
GDP	Trillion $	2.9	5.3	9.5	16.0
GDP growth rate	%	6.0	6.5	5.7	5.0
Population	Billion	1.38	1.50	1.59	1.64
Urbanization	%	37	42	46	50

Source: 1. Shaojie song 2. Haiyeng Lin 3. Peter Sherman 4. Xi Yang 5. Shi Chen 6. Xi lu 7. Tianguang Lu 8. Xinyu Chen 9. Michael B. McElroy.

Table 11.2 Projected road transport vehicles stock (in units of million vehicles).

	2015	2020	2025	2030	2035	2040	2045	2050
Two-wheelers	169	198	233	294	372	471	635	856
Three-wheelers	6	9	12	17	24	35	48	66
Four-wheelers	36	47	61	76	95	119	169	241
Buses	3	3	3	3	4	4	6	7
Trucks	4	7	12	15	18	22	25	28

Source: 1. Shaojie song 2. Haiyeng Lin 3. Peter Sherman 4. Xi Yang 5. Shi Chen 6. Xi lu 7. Tianguang Lu 8. Xinyu Chen 9. Michael B. McElroy.

passenger transport service are per capita income and responses to increasing trends in urbanization and population. The vehicle stock for the base year (2015) selected by Song et al.[8] in their seminal study, adopted also for the present analysis, was taken from the *India Statistical Yearbook* published in 2018 by the Ministry of Road Transport and Highways.[9] Data on fuel economy, mileage per vehicle, and load factors were adopted from a variety of sources including the IEA and studies by Paladugula et al.[10] and Prakash and Habib.[11] The vehicle stocks for 2015 and projections for 2050 were based on the study by NITI Aayog, results for which are summarized in Table 11.2.

In 2015, the transportation sector was fuelled almost entirely by petroleum products (diesel, gasoline, and, in the case of aviation, jet fuel), consuming the equivalent of approximately half a billion barrels of oil, rising to approximately 900 million barrels (121.5 Mt) in 2021–2022.[12] As described in Chapter 8, the government is encouraging Indian auto makers to test the market for electric vehicles and hybrids that could use a combination of liquid fuels and electricity. It has begun also to divert some of India's growing surpluses of cereals into the production of ethanol, setting asperationally high targets for its addition to gasoline by 2025 and 2030. But these are short-term expedients. The unspoken hope within the government is that by the 2050, the target date selected here, the bulk of the Indian transportation sector, with the possible exceptions of aviation, shipping, and heavy-duty road haulage, will be fuelled by electricity, with a potential additional contribution from products derived from biomass. It is conceivable, though, given the paucity of investible resources and demands to meet other pressing developmental needs (expansion of opportunities for employment for example) that this objective may not be realized until later. For present purposes, following Song et al.[13], we assume that light-duty vehicles (two-, three-, and four-wheelers) will be propelled mainly by battery-powered electricity in 2050, requiring an annual source of a little more than 1,000 TWh. Heavy-duty trucks and buses will most probably continue to be fuelled by oil and oil derivatives, or potentially, assuming that road safety considerations can be satisfactorily addressed, by carbon-free green hydrogen. The former would require an annual source of 1.3 billion barrels of oil equivalent, the latter 32 Mt of hydrogen.

Energy in the residential, commercial, and agricultural sectors

The demand for energy in the residential sector includes requirements for heating, cooling, lighting, hot water, cooking, and appliances. It is satisfied currently by a combination of fossil fuels (primarily kerosene and liquefied petroleum gas), biomass, and electricity. Demand for energy in the commercial sector involves primarily requirements for lighting, air conditioning, and appliances. The demand for air conditioning is already significant, growing rapidly in response to the hot and humid conditions experienced recently, a trend that is likely to become even more important in light of the warmer conditions projected for the future, amplified additionally by anticipations for increased affluence and urbanization. Our projections of the increasing demand for energy in the

Table 11.3 Final energy service demand prediction in the residential, commercial, and agriculture sectors (in petajoules).

	2015	2020	2025	2030	2035	2040	2045	2050
Residential	3989	4293	4747	5422	6418	7787	9638	12189
Commercial	635	777	1002	1236	1541	1924	2391	2973
Agriculture	1037	1074	1113	1154	1375	1638	1952	2326

residential and commercial sectors and the growing requirements for agriculture are summarized in Table 11.3. The demand for energy in the agricultural sector is assumed to be driven primarily by the growth of GDP. The results in Table11.3 are presented in units of 10^{15} joules [referred to as petajoules (PJ)]. A petajoule is a unit of energy equivalent to 9.48×10^{11} British thermal units (Btu), 2.78×10^{8} KWh, or 0.278 TWh. The data in Table 11.3 indicate that between 2015 and 2030 the demand for energy in the residential, commercial, and agricultural sectors will increase by factors of 3.1, 4.7, and 2.2 respectively.

Energy in the industrial sector

Predictions for energy requirements in the industrial sector depend on the projected mix of products. A summary of demand generated by eight of the most energy-intensive products, cement, iron and steel, aluminium, caustic soda, soda, paper, and textiles, is presented in Table 11.4a in units of million tonnes and in Table 11.4b in units of petajoules. Interactions among sectors have been taken into account in developing this database. Taking iron and steel for example, the demand in this case involves mainly three components: residential and commercial buildings, household appliances, and vehicles in the transportation sector. The data in Table 11.4a suggest that the aggregated demand for iron and steel in these sectors will rise from 95.5 MT in the 2015 base year to 222.5 MT in 2030, reaching 364.6 MT in 2050, corresponding to an enhancement in related energy demand over these years from 1,665.5 PJ (2015), to 2,875.2 PJ (2030), to 6,381.3 PJ (2050), consistent with projections by NITI Aayog[14] and UNEP.[15]

Table 11.4a Key industrial product demand projections (in million tonnes).

	2015	2020	2025	2030	2035	2040	2045	2050
Cement	283.5	349.0	483.0	710.0	815.1	935.8	1074.4	1250.0
Caustic soda	2.5	5.0	6.7	9.0	10.8	12.6	14.3	16.3
Soda	2.6	5.1	6.8	9.1	10.8	12.7	14.4	16.4
Fertilizer	41.3	43.6	46.3	48.9	51.6	53.5	55.7	57.9
Iron and steel	95.5	126.6	167.8	222.5	251.7	284.8	322.2	364.6
Aluminium	2.4	2.9	3.9	5.2	5.9	6.6	7.5	8.5
Paper	17.0	22.5	29.9	39.6	43.7	48.3	53.3	58.8
Textiles	68.4	79.3	91.9	106.6	117.0	128.6	141.2	155.3

Source: 1. Shaojie song 2. Haiyeng Lin 3. Peter Sherman 4. Xi Yang 5. Shi Chen 6. Xi lu 7. Tianguang Lu 8. Xinyu Chen 9. Michael B. McElroy.

Table 11.4b Key industrial product demand projections (in petajoules).

	2015	2020	2025	2030	2035	2040	2045	2050
Cement	844.8	964.6	1388.5	1965.6	2171.8	2453.8	2772.3	3179.4
Caustic soda	36.1	71.3	92.5	119.8	139.2	156.5	173.2	192.8
Soda	37.5	72.8	93.8	121.2	139.2	157.8	174.4	193.9
Fertilizer	678.1	710.8	733.3	751.7	760.3	755.6	755.9	760.5
Iron and steel	1665.5	2002.8	2361.8	2875.2	3581.6	4396.9	5486.8	6381.3
Aluminium	50.3	55.8	63.6	74.0	86.4	103.8	120.9	135.3
Paper	463.7	581.3	739.3	936.8	993.6	1064.2	1116.5	1173.4
Textiles	146.8	166.0	187.9	212.9	228.8	246.1	264.9	285.7
Total (PJ)	3922.8	4625.4	5660.7	7057.2	8100.9	9334.7	10864.9	12302.3
Total (TWh)	1089.7	1284.8	1572.4	1960.3	2250.3	2593.0	3018.0	3417.3

Source: 1. Shaojie song 2. Haiyeng Lin 3. Peter Sherman 4. Xi Yang 5. Shi Chen 6. Xi lu 7. Tianguang Lu 8. Xinyu Chen 9. Michael B. McElroy.

The COVID-19 pandemic, however, put a brake on this growth. Iron and steel production remained at 100.2 MT in 2020 against the predicted 126 MT. But production rebounded to 118.2 MT in 2021.[16] The 9% annual growth required to meet NITI Aayog's projection of 222.5 MT in 2030 is therefore ambitious but still achievable. The same could be true for the other main industrial consumers of energy—cement, fertilizer, and paper. Data on energy intensity and projected improvements in energy efficiency over time are taken from UNEP.[17]

Summing up

The annual demand for electricity is expected to grow from 947 TWh in 2015 to 8,323 TWh in 2050. Contributions to the time-evolving demand for the five primary components of the economy out to 2050 are summarized in Table 11.5. The electrification rates for the residential, commercial, and agricultural sectors are taken as 100% in 2050. The electrification rate for the industrial sector is taken as 34% in 2050, consistent with the analysis by TERI/Shell[18], with the bulk of the energy demand in this case, supplied by green hydrogen with, most likely, though, a continuing, hopefully minor, role for fossil fuels.

Electricity is expected to make a major contribution to the demand for energy by light-duty vehicles in the transportation sector, with a potentially important contribution at least in the immediate future, as discussed in Chapter 12, from fuels developed from biomass. Particularly urgent will be the need for development of a zero carbon fuel for aviation, for which there is at present no viable alternative to fossil fuels, although an initiative in Reno Nevada, involving gasification of municipal waste followed by conversion of the syngas product using the Fisher–Tropsch process, shows promise. Biomass could also make an important contribution to the demand for energy in a number of industrial sectors. In steel, for example, which is by far the largest consumer of heat, coking coal, which is needed to produce pig iron in blast furnaces, could be replaced by biochar briquettes formed by gasifying biomass. This option is being seriously considered by Canada.[19] Most other countries, however,

Table 11.5 Projected annual national-level electricity consumption by sector (in terawatt hours).

	2015	**2050**
Industry	395	2103
Residential	239	3386
Commercial	119	826
Transport	22	1361
Agriculture	173	646
Total	948	8322

Source: 1. Shaojie song 2. Haiyeng Lin 3. Peter Sherman 4. Xi Yang 5. Shi Chen 6. Xi lu 7. Tianguang Lu 8. Xinyu Chen 9. Michael B. McElroy.

including India, have shown little interest in following Canada's example. The main hurdle is the high cost of collecting the biomass required to form the biochar. Possible ways around this hurdle are discussed in Chapters 12 and 15.

Future hourly demand for electricity

India's most urgent need is to break the chokehold of coal on the generation of electricity in the residential, commercial, and industrial sectors, together with the need to cut back on the demand for petroleum in the transportation sector. In what follows we explore four models for the 2050 Indian energy demand. All four models consider requirements for power on an hourly basis over the course of a year in five regions of the country (East, North, South, West, and North-east). The starting point involves regional hourly demands for 2019 reported by the Power System Operation Corporation of the Indian Ministry of Power.[20] Hourly variations of temperature with time in 2019 were used to compute cooling and heating degree days (CDDs and HDDs). CDDs and HDDs reflect the difference between the daily mean value of the temperature and 65 °F (18.3 °C), taken to indicate the threshold distinguishing the need for either heating or cooling over the course of a day. If the daily mean temperature is above 65 °F, we subtract 65 °F from the mean and this defines the contribution to CDDs. Similarly, if the daily mean is below 65 °F, we subtract the mean from 65 °F and this defines the contribution to HDDs. Hourly variations of temperature with time daily in 2050 are assumed to follow the pattern observed in 2019. Population weighted CDDs and HDDs were used to represent the regional daily shares of power for annual heating and cooling. Demand for electricity in the transportation sector was allocated first on a regional basis according to the current geographical distribution of registered vehicles,[21] downscaled then to hourly resolution, ignoring possible variations in daily apportionment.

The detailed procedures used to project the overall hourly and regional demands for power in 2050 are described by Song et al.[22] Results for the individual regions are presented in Figure 11.1. The panels to the left indicate the variability projected on a daily basis over the course of the year. The panels to the right reflect the variability expected on a 24-hour

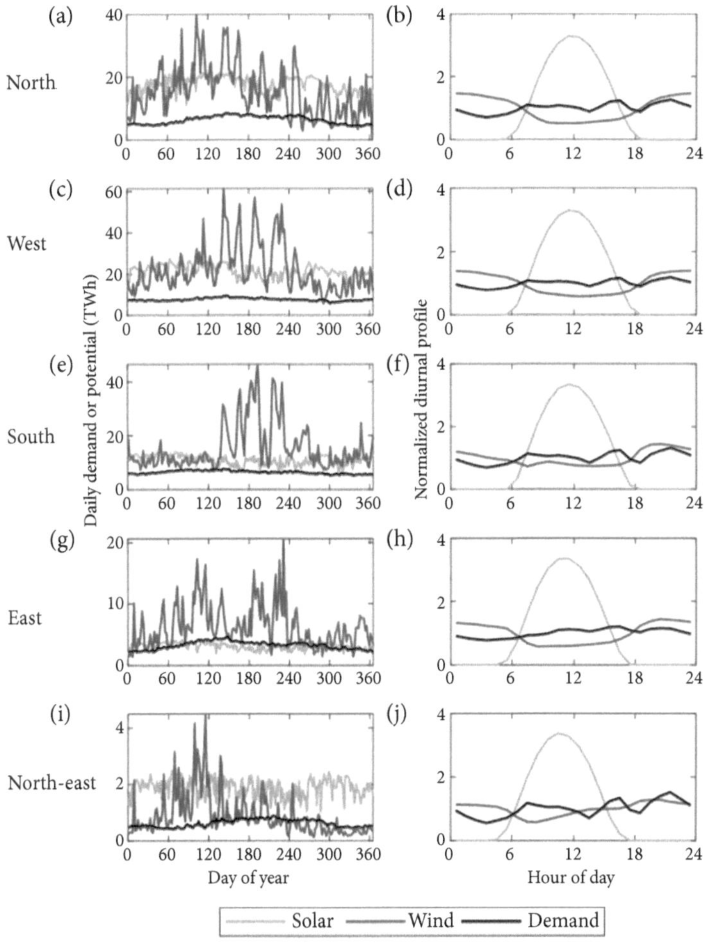

Figure 11.1 Solar and wind power potential and 2050 demand for the five regions.

Note: Left-side panels (a, c, e, g, and i) show daily variability and right-side panels (b, d, f, h, and j) indicate diurnal profiles normalized by 24-hour average values (y-axis unit free).

Source: Created by the authors using data by Song et al.

basis for each of these regions. Figure 11.1 includes also, for reference, the potential sources of power that could be derived from wind and solar following the discussion in Chapter 9. Clearly, the combination of wind and solar is more than sufficient to meet the projected demand for power on both a daily basis and over the course of the year.

This is needless to say a measure of potential. It does not, however, take into account the willingness of the people who own or live on the land to give up some or all of their property to accommodate the installation of the wind and/or solar resources considered here. An underlying challenge, is to accommodate for the intrinsic variability of these sources not only regionally and seasonally but also diurnally. There will be times, for example, when the potential supply of power from either solar or wind will be more than sufficient to meet demand, offset by times when the supply may be insufficient. The discrepancy can be resolved by investments in energy storage: batteries and pumped hydro to accommodate variability in the short term; and hydrogen to address variability in the longer term, as discussed in Chapter 10. Investments in the interregional transmission network can also play a role, coupling regions of power surplus to regions of deficit. Following the approach outlined by Song et al.[23], we consider four scenarios to meet India's projected 2050 energy future. The objective is to identify the least-cost options to meet demand subject to the constraints implicit in the different models.

Four model scenarios for 2050

The baseline scenario, S1, effectively represents business as usual. It is dominated by fossil fuels. The capacity of renewable power is set at 450 GW for 2050, consistent with the target for 2030 announced by the Indian government in 2016.[24] The additional wind and solar energy projected to be developed between 2022 and 2030 will meet a portion of the demand for electricity in the transportation sector (excluding buses and long-distance trucks), in agriculture (pumping and tractors), and in servicing commercial and residential buildings. Capacities for nuclear and hydro are assumed to be fixed at current levels. Reductions in costs for renewable power systems, thermal plants, transmission lines, and storage are assumed to follow the moderate trend identified in the National Renewable Energy Laboratory (NREL) Annual Technology Baseline (ATB) database.[25] Costs for coal and natural gas in 2050 are taken as $3.6 per million British thermal units (MMBtu) and $7.65 per MMBtu respectively, consistent with assumptions in the earlier paper by Lu et al.[26] Oil costs for the same year are estimated at $70 per barrel.

Emission factors for coal- and natural gas-fired units were taken from Sharma et al.[27] Capital costs for solar and wind corresponded to \$618 per kW and \$1,055 per kW respectively in 2019 (NREL[28]). Cost reductions for these technologies between 2020 and 2050 are taken as 49% and 39% respectively, consistent with the moderate trend defined by the NREL.[29] This would imply capital costs for solar and wind in 2050 of \$315 per kW and \$644 per kW respectively.

Scenario 2 (S2) explores the possibility for wind and solar energy to provide the level of electricity contemplated in S1 without the need to include the contributions from fossil fuels assumed in the latter. Scenarios 3 and 4 (S3 and S4) assume that renewable sources account for 100% of power production, addressing specifically opportunities and benefits for integration of green hydrogen into the energy mix. Hydrogen in S3 is applied only in the power sector, providing an opportunity to compensate for shortfalls in supply that would arise otherwise given the variability of the wind and solar source. S4 considers a more expansive application of green hydrogen with the purpose of facilitating decarbonization of otherwise hard to abate sectors in both industry and transport, while at the same time eliminating the current dependence on grey hydrogen.

Results

As indicated earlier, the objective for each scenario is to identify the least-cost approach to meet the stated goals of the scenario. Selected expansion of the interregional power transmission network plays an important role in meeting this objective. Figure 11.2 presents a summary of the sources of energy that will need to be deployed in the different regions by 2050. Figure 11.3 summarizes the exchange of energy projected between the different regions associated with the expanded interregional transmission system. The capacity of the required transmission network ranges from a low of 173 GW in S1 to a high of 1,355 GW in S4.

The link between the South and West is particularly significant for models S2, S3, and S4. The link between the North and West is most important for S1. Underscoring the importance of investments in the transmission network, close to 30% of the energy consumed in the West under S2 is supplied from the South. In S2, S3, and S4, wind in the South and

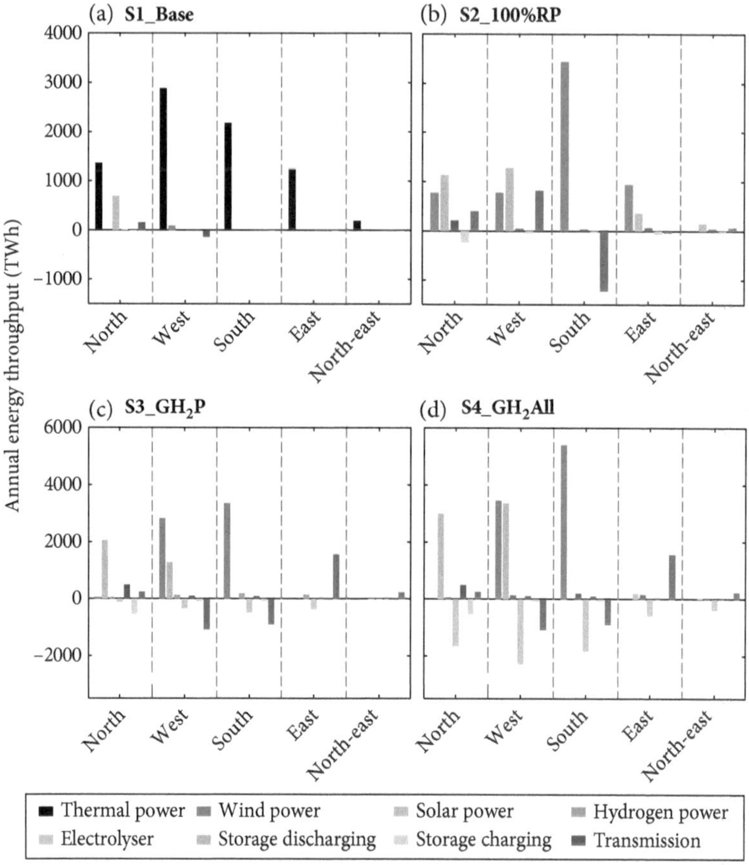

Figure 11.2 Regional distribution of annual energy throughput for different technologies for 2050 inferred using the optimization model. Negative transmission values identify exports; positive values identify imports.

Note: Curtailed solar and wind power are not included in the figure. Data identified as transmission refer to interregional transfers of power over the transmission network.

West and solar in the North and West are both significant sources of electricity, consistent with the potential for the supplies of renewable power available in these regions as indicated by the data presented in Chapter 9. Under the hydrogen pathways, S3 and S4, the bulk of the electricity consumed in the East and North-east is derived from other regions and used in large part for electrolysis. Overall, electricity generated from wind is greater than that from the solar source, reflecting the higher capacity factors associated with the former (see discussion in Chapter 7).

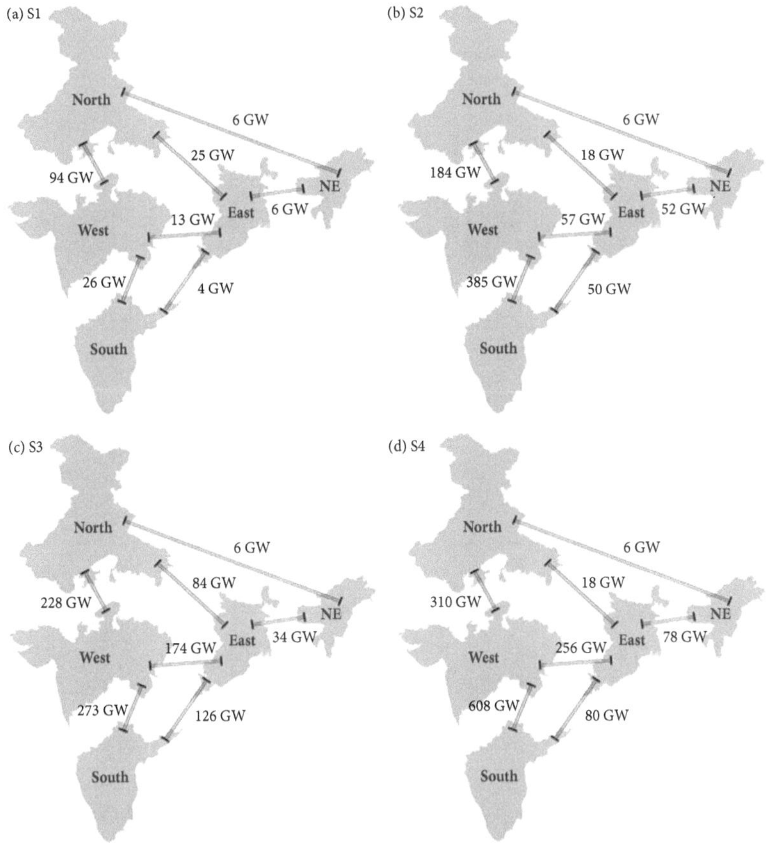

Figure 11.3 Exchange of energy between the different regions associated with the extended transmission network projected for the four scenarios under the cost minimization methodology described by Song et al.

Note: Not to scale; for representational purposes only.

Source: Created by the authors using data by Song et al.

The diurnal and seasonal patterns for availability of wind and solar power can have an important impact on hourly system operations throughout the year, particularly for the scenarios that depend on large capacities of renewables. Hourly power balances for representative days over the four quarters of the year are presented in Figure 11.4. The days selected—1 January, 1 April, 1 July, and 1 October—correspond broadly to the midpoints of India's brief winter season (1 December to 28 February), the summer season (1 March to 31 May), the monsoon season (1 June to 31 August), and the post monsoon autumn period

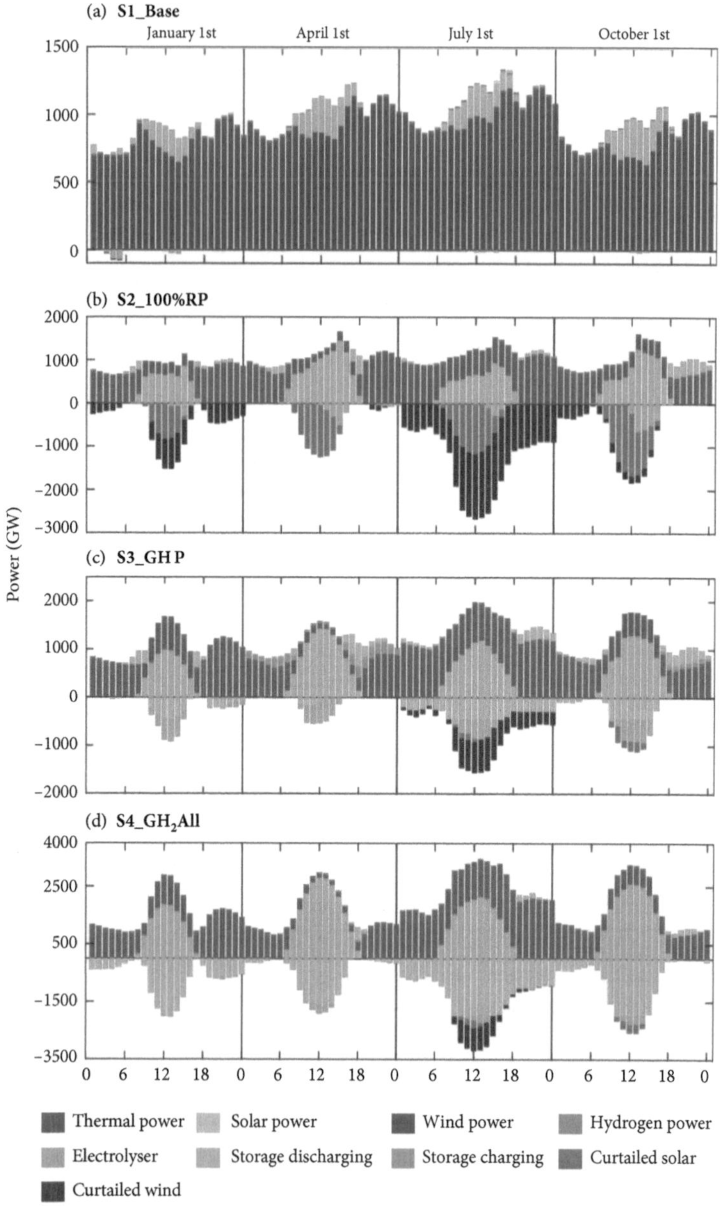

Figure 11.4 Seasonal hourly power flow (in GW).

Note: Countrywide hourly power flows are defined for seasonally representative days in 2050 based on the model simulation for the selected four scenarios. Positive values refer to power generation; negative values refer to storage of power, curtailed electricity, and use of power to produce hydrogen.

(1 September to 30 November). In S1, thermal power remains the dominant source of electricity for all four seasons, with relatively minor contributions (8.7%) from wind and solar. In S2, wind and solar are the dominant sources of electricity, with solar a significant contributor in afternoons, with wind playing a major role throughout the day, particularly in evenings. However, with the adoption of these intermittent sources of power without compatible options for energy storage, large amounts of both wind- and solar-generated electricity are curtailed, most notably during the monsoon season when curtailment could correspond to more than 50% of the potential source. This underscores the importance of the integration of hydrogen technologies in S3 and S4 which allows for much of the power that would otherwise be curtailed to be used for electrolysis. The national average annual renewable power curtailment is reduced as a consequence from 35.7% in S2 to 7.3% and 3.3% in S3 and S4 respectively.

The installed capacity for hydrogen-producing electrolysis amounts to 463 GW and 1,982 GW in S3 and S4 respectively. Deploying electrolysis fully will raise the annual average capacity factors for the two scenarios to 34% and 39.4% in S3 and S4 respectively. The marginal price for production of hydrogen in S4 declines to an estimated $1.3 per kg. The price for hydrogen is incorporated in the price for electricity in S3.

National annual costs for the supply of energy in the four models are summarized in Figure 11.5. The contribution labelled 'thermal power' in S1 refers to the costs for fuel, amortized capital, and operational and maintenance expenses for coal- and gas-powered plants. The costs labelled 'hydrogen related', which are prominent in S3 and S4, include the capital and operational expenses for the electrolysers, the costs for the hydrogen turbines employed for power generation in S3, and the costs for storing hydrogen. 'Fossil fuel other' indicates the costs for coal, oil, and natural gas consumed in sectors other than power generation. 'Others' refers to costs for amortized capital and operational and maintenance expenses associated with hydro and nuclear power. The total costs of energy supply range from a low of $493 billion a year in S3 to a high of $596 billion a year in S2. The high cost in S2 reflects the large quantity of renewable power that has to be curtailed in that scenario. Figure 11.5 includes also estimates for the annual emissions of CO_2 associated with the

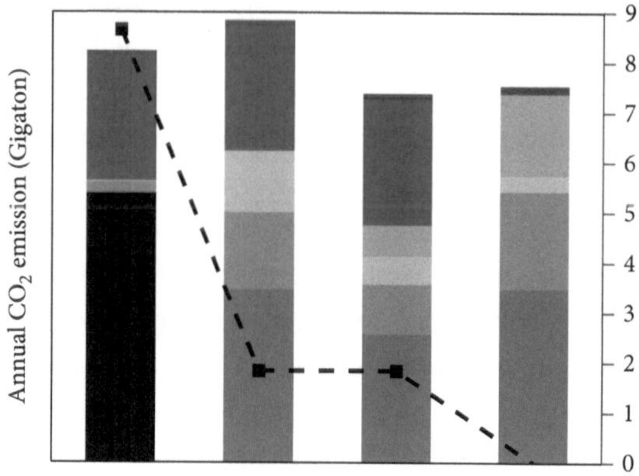

Figure 11.5 National energy supply cost.

different scenarios, which range from 8.64 Gt annually in S1, decreasing to 1.83 Gt in S2, rising to 1.83 GT in S3, falling to zero in S4.

Concluding remarks

The preceding discussion considered four routes by which India could meet its projected demand for energy in 2050. S4 offered the most ambitious scenario to address this objective. It envisages a major investment in green hydrogen complementing an important commitment to wind- and solar-generated electricity, taking advantage of the intrinsic variability of these sources. Hydrogen in the current Indian economy is deployed mainly in the oil refinery sector and as an input to the manufacture of nitrogen fertilizer. The present source of hydrogen in these applications is grey, produced largely from natural gas with associated release of approximately 10 kg of CO_2 for every kg product of hydrogen. The green hydrogen product in contrast is carbon free so long as the electricity used to produce it is derived from a carbon-free source such as wind or solar. Green hydrogen is deployed in S4 not only to replace current industrial applications of the grey version but also as a substitute for coking coal in the iron and steel industry, as a fuel for portions of the transportation sector, and potentially as input to a variety of other applications where it can take the place of carbon-emitting sources of fossil energy (see Chapter 9 for a more detailed discussion). The analysis presented here suggests that the scenario envisaged in S4 could provide an economically viable alternative to the business-as-usual alternative S1 for India to meet its projected future energy needs while at the same time minimizing its emissions of CO_2.

We recognize that the options considered here are not the only possibilities available for India. It would be possible, for example, for the country to elect to use its abundant sources of urban and rural organic waste to contribute to its energy future. At present, much of this material is incinerated, with serious negative impacts not only on air quality but also on public health.

Processing this material could provide a valuable source of chemicals that could be deployed in a variety of sectors of its economy, ranging from domestic, to commercial, to industrial, to transportation. In particular, collection and processing of agricultural waste could offer a much-needed second source of income and employment in rural areas, while at the same time providing an important boost to rural economies.

The analysis developed in this chapter placed an emphasis on minimizing the costs involved in meeting India's projected future energy needs subject to the specific constraints imposed in the different scenarios. But costs and benefits to investors are not the only factors that need to be considered in charting India's path to its future. Planning should also take account of the impact on the lives and welfare of the nation's citizens. For good reasons, under a variety of circumstances, social factors might outweigh economic considerations in determining the precise trajectory the nation should follow on its path to decarbonization. These factors are discussed in greater detail in the chapters that follow.

Several sectors were not explicitly treated in the present study, including for example aviation and shipping. As mentioned in Chapter 10, biomass sources could be applied to produce jet fuel, and ammonia produced from green hydrogen could provide fuel for shipping. A more comprehensive analysis should consider also opportunities for sequestration of carbon through purposeful investments in managed ecosystems, in forests and soils for example. And, finally, we would note that the economy as a whole is likely to benefit significantly from the transition to the domestic sources of energy such as wind, solar, and biomass emphasized here, a shift that would eliminate the need to expend major sources of funds to support imports of expensive commodities such as oil and natural gas and even high-quality coking coal, as is the situation at the moment. Improvements in the nation's balance of payments could free up scarce sources of foreign reserves that could be used productively elsewhere in the domestic economy, including the funding required for development of the renewable energy alternatives emphasized here.

Notes and References

1. https://www.iea.org/reports/india-energy-outlook-2021/energy-in-india-today (viewed 23/01/2025).
2. NITI Aayog: Energizing India, 2017. https://niti.gov.in/sites/default/files/energy/Energising-India.pdf (viewed 3/8/2022).
3. India GDP Growth Rate 1961–2022. Macro trends. https://www.macrotrends.net/countries/IND/india/gdp-growth-rate (viewed August 3, 2022).
4. Song, S., et al. (2022). Deep decarbonization of the Indian economy: 2050 prospects for wind, solar, and green hydrogen. *iScience* 25(6).
5. IEA: India Energy Outlook 2021. https://www.iea.org/reports/india-energy-outlook-2021 (viewed 5/8/2022).
6. United Nations, Department of Economic and Social Affairs, Population Division (2019). World Population Prospects 2019: Highlights. https://www.un.org/en/desa/world-population-prospects-2019-highlights (viewed 5/8/2022).
7. NITI Aayog: Energizing India, 2017. https://niti.gov.in/sites/default/files/energy/Energising-India.pdf (viewed 3/8/2022).
8. Song, S., et al. (2022). Deep decarbonization of the Indian economy: 2050 prospects for wind, solar, and green hydrogen. *iScience* 25(6).
9. Ministry of Road Transport and Highways (2018). *Road Transport Year Book (2015–16)*. Government of India.
10. Paladugula, A.L., et al. (2018). A multi-model assessment of energy and emissions for India's transportation sector through 2050. *Energy Policy* 116:10–18. https://doi.org/10.1016/j.enpol.2018.01.037 (viewed 9/8/2022).
11. Prakash, J. and Habib, G. (2018). A technology-based mass emission factors of gases and aerosol precursor and spatial distribution of emissions from on-road transport sector in India. *Atmospheric Environment* 180:192–205.
12. TERI/Shell (2021). India: Transforming to a Net-Zero Emissions Energy System. www.shell.in/Indiasketch (viewed 9/8/2022).
13. Song, S., et al. (2022). Deep decarbonization of the Indian economy: 2050 prospects for wind, solar, and green hydrogen. *iScience* 25(6).
14. NITI Aayog: Energizing India, 2017. https://niti.gov.in/sites/default/files/energy/Energising-India.pdf (viewed 3/8/2022).
15. Vishwanathan, S.S., et al. (2017). Enhancing Energy Efficiency in India: Assessment of Sectoral Potentials. Copenhagen Centre on Energy Efficiency, UNEP DTU Partnership.

16. Ministry of Steel, Government of India: Make in India. https://steel.gov.in/make-india#:~:text = Introduction,period%20last%20year%20(CPLY) (viewed 11/8/2022).
17. Vishwanathan, S.S., et al. (2017). Enhancing Energy Efficiency in India: Assessment of Sectoral Potentials. Copenhagen Center on Energy Efficiency, UNEP DTU Partnership.
18. TERI-Shell, Commodity Insights. https://www.spglobal.com/commodityinsights/en/market-insights/latest-news/oil/031021-indias-fy2021-2022-transport-fuel-consumption-to-remain-expansionary-as-covid-19-fears-ease (viewed 12/8/2022).
19. Peter Warrian and Homeira Afshar: Innovation and de-carbonization in the Canadian steel industry. chrome-extension://efaidnbmnnnibpcajpcglclefindmkaj/https://munkschool.utoronto.ca/media/2648/download?inline=/download. Innovation Policy Lab Working Paper Series 2021–23 (viewed 23/01/2025).
20. https://powermin.gov.in/en/content/power-sector-glance-all-india
21. Ministry of Road Transport and Highways (2018). *Road Transport Year Book (2016–17)*. Government of India.
22. Song, S., et al. (2022). Deep decarbonization of the Indian economy: 2050 prospects for wind, solar, and green hydrogen. *iScience* 25(6).
23. Ibid.
24. Government of India (2016). India's Intended Nationally Determined Contribution: Working towards climate justice. https://www4.unfccc.int/sites/ndcstaging/PublishedDocuments/India%20First/INDIA%20INDC%20TO%20UNFCCC.pdf (viewed 17/8/2022).
25. NREL (2020). 2020 Annual Technology Baseline. https://atb.nrel.gov/ (viewed 23/8/2022).
26. Tianguang, Lu, et al. (2020). India's potential for integrating solar and on- and offshore wind power into its energy system. *Nature Communications* 11:1–10.
27. Sharma, B., et al. (2018). CO_2 Baseline Database for the Indian Power Sector. Government of India, Ministry of Power, Central Electricity Authority.
28. NREL (2020). 2020 Annual Technology Baseline. https://atb.nrel.gov/ (viewed 23/8/2022).
29. Ibid.

12

Biomass Gasification Can Complete the Journey to a Net Zero World

In the preceding chapters we described the potential of wind and solar energy to meet India's future requirements for electricity and showed that this comfortably exceeds the demand projected for 2050. We suggested several ways to utilize this surplus, of which the most important involved the use of green hydrogen as a source of high-temperature process heat to be applied in the hard to abate sectors of industry, notably steel, cement, petrochemicals, and fertilizers. But neither solar nor wind energy can replace liquid fuels in the transport sector directly. Doing so would require a complete shift out of liquid fuels to battery-stored electricity, not only for the 1.45 billion cars on the roads globally in 2022 but also for the billion or more projected to reach the market over the succeeding 30 years, even if the growth of car ownership, which has averaged 4.49% a year since 2001, is halved over this period.[1] Since the average life of an automobile in most low-income countries up to 20 years or more in most low-income countries, and 14 years even in the US, replacing close to three billion liquid-fuelled automobiles with EVs and creating the infrastructure to support them in less than the few decades the world has if it is to limit the increase in global average surface temperature to less than 1.5 °C relative to the pre-industrial standard, as recommended by the IPCC[2], will pose a major challenge. It will be especially difficult for developing countries where GDP growth and car ownership are rising most rapidly, where investible resources are scarce, and where governments have to meet a host of other more pressing demands, such as health, education, and employment. NITI Aayog[3] projects that the total number of vehicles on the road in India is likely to grow from the present level of 296 million to more than a billion by 2050. Realistically, if we are to reach the zero carbon future targeted for 2050,

Sun, Wind, and Biomass. Prem Shankar Jha and Michael B. McElroy, Oxford University Press.
 DOI: 10.1093/9780198944768.003.0012

it may be necessary to rely on a mix of EVs and hybrid vehicles using a combination of zero carbon electricity and zero carbon liquid fuels derived from biomass rather than the pure battery-operated alternative. This could ensure a future road-based transportation system for India, and potentially also for the world, that could be either carbon neutral or potentially even carbon negative.

The most immediate need for biofuels, however, is in aviation. The fact that there is no alternative to biofuels in this area was accepted by the Biden administration. At the Glasgow climate change conference, Biden's Transportation Secretary, Pete Buttigieg, released the *Aviation Climate Action Plan*, prepared by the US Federal Aviation Administration (FAA).[4] The Plan's endorsement of biofuels was unambiguous:

> Sustainable fuels produced from renewable and waste feedstocks can provide the greatest impact in our effort to reduce aviation's greenhouse gas emissions. Such fuels will be critical to the aviation industry's ability to meet the net-zero emissions goal and they have the potential, on a lifecycle basis, to slash emissions by up to 100%. Sustainable aviation fuels can be used in today's fleet of aircraft without modification and can be produced from a wide range of feedstocks, including wastes, residues, biomass, sugar, oils and gaseous sources of carbon.

Biomass can be converted into transport fuels or associated chemicals through either fermentation or gasification. The first is a biological process currently in wide use to produce ethanol. The second is a thermochemical process that can be made to yield almost any form of transport fuel, from methanol, through high-speed diesel, to aviation turbine fuel. It could be used also to produce heavy carbon-rich gases such as DME, which could replace liquefied petroleum gas (LPG) both as a transport fuel and as a cooking fuel. The two key reactions that make this possible are Fischer–Tropsch synthesis and the water gas shift reaction. Both have been around for more than 100 years. They were used by Germany in the 1940s to produce diesel and aviation fuel to fight the Second World War and by South Africa in the 1980 to defeat a UN-imposed embargo on the sale of crude oil. Coal provided the feedstock used in both these cases.

Gasification and the Fischer–Tropsch synthesis: the basics

Gasification differs from combustion in that the fuel is exposed in the former case to a limited and carefully calibrated supply of either air or oxygen. While combustion generates only CO_2, the product of gasification involves a mix of carbon monoxide, hydrogen, and CO_2 together, especially at lower gasification temperatures, with methane and other impurities.

This mixture, minus the methane and impurities, is referred to as synthesis gas or syngas and is the basic building block for much of the vast synthetics and petrochemical industries in the world today. Interest in using biomass in place of fossil fuels developed only after the first oil price shock in the 1970s. But although carbon in the form of lignin makes up a major portion of most forms of biomass, the wide diversity in its chemical composition and moisture content makes it necessary to develop gasifier designs capable of coping with a variety of different properties of the potential biomass feedstock. By the late 1980s, biomass gasifiers that had found ways around this variability had been developed in laboratories and proved in pilot plants across the world. For a description of the variety of gasifiers on the market today, classified variously as Updraft Fixed Bed (UFB), Downdraft Fixed Bed (DFB), Bubbling Fluid Bed (BFB), Circulating Fluidized Bed (CFB), Entrained Flow Gasifiers (EFG), and Plasma Gasifiers (PG), the reader is referred to the review by Quaak et al.[5]

The choice of a particular gasifier and the details of its implementation depend on the targeted plan for utilization of the product synthesis gas. In what follows we focus exclusively on the use of biomass as feedstock for the production of transport fuels. A definitive study by Specht et al.[6] described the complexity of the challenge confronted in meeting this objective. The researchers explored three different ways of producing methanol from wood chips. All three used steam as an input to the gasifier, with oxygen supplied by a pressure swing adsorption (PSA) plant in model 1. And from electrolysis supplying oxygen and hydrogen in models 2 and 3. All three yielded fuel gas, consisting of Carbon Monoxide and Hydrogen, but also significantly differing amounts of CO_2

which had to be removed before the raw gas became appropriate for the synthesis into methanol or other synthetic fuels. In Model 1, 61% of the CO_2 had to be vented into the atmosphere, resulting in the conversion of as little as 44% of the carbon in the biomass into methanol.

In Model 2 therefore Specht and his colleagues used an electrolyser to supply not only the needed oxygen but an important amount of hydrogen that could react with the excess of CO_2 being created by the gasifier. This higher concentration of hydrogen enabled a large proportion of the CO_2 in the fuel gas to also be converted into methanol because Carbon Dioxide requires 3 molecules of hydrogen, along with Oxygen, to yield methanol while Carbon Monoxide (CO) requires only 2 molecules of hydrogen to do so.

In Model 3, they therefore used a larger electrolyser to produce not only the needed oxygen but also sufficient additional hydrogen to convert close to all of the CO_2 into methanol. The addition of the extra hydrogen allowed nearly all of the CO_2 to also be converted into Methanol.

The experience developed in the Specht et al.[7] study underscores the importance of a careful analysis of the steps involved in the conversion of biomass to fuel. It is not simply a matter of adding the biomass to a gasifier, adjusting the CO/H_2 ratio using the water gas shift reaction, and transferring the product to the Fischer–Tropsch reactor. All of the intermediate steps need to be carefully considered, paying careful attention to the different properties of the initial biomass fuel, the precise path for gasification, and the nature of the catalysts required by the ultimate Fischer–Tropsch for the processing of the syngas into biofuels.

In the US, the first commercial venture using biomass as the feedstock for producing transport fuels was announced in 2018 by a company named Sierra Biofuels in May 2018. Its purpose was to produce aviation turbine fuel (ATF) from Municipal solid, in Reno, Nevada, Its purpose was to turn out as much as 11 million gallons of 'green' aviation fuel and 'green' renewable diesel every year. The MSW, instead of going to a landfill, was to be redirected to a processing facility operated by the company, where metals and inorganic materials would be separated from the organic component of the waste, which would then be fed into a two-stage gasifier from where a cleaned and dried fuel gas would be fed into a biorefinery where it would be converted into ATF and possibly diesel, through a well known and widely used process known as the

Fischer–Tropsch synthesis. The company began operations at its plant in Reno in 2022, processing 175,000 tonnes of municipal waste, to produce up to 11 million gallons of sustainable aviation fuel (SAF) or renewable diesel, annually.[8]

The company had plans to replicate such facilities across the US with the intention of producing up to 400 million gallons of synthetic crude annually. But at a fairly early stage it ran into teething troubles—notably an accumulation of nitric acid in its upper pyrolysis chamber—that forced it to halt operations till the acid was cleared and its cause diagnosed. To finance the work involved it asked for a loan from the Federal government under the Biden-Buttigeig Plan for Infrastructure development, but no sooner had it done so than a group of US senators, of different party affiliations demanded that the loan be denied because the grants programme was not intended for 'non-disposable' waste. Despite the fact that only a part of the Municipal Solid Waste was non-disposable, and this was not the part that the company would be working with, the Senators prevailed and the company went into Liquidation. The viability of the technology that Sierra biofuels intended to use has not, however been questioned. A number of Biofuel projects have begun in the United States and other countries include projects to produce renewable diesel, biojet kerosene, and ethanol. These projects use a variety of feedstocks, including maize stover, forestry residues, and municipal solid waste.[9]

In what follows we plan to discuss the opportunities for quantification and exploitation of India's important sources of biomass. We propose to emphasize the prospects for production and use of methanol in the road-based transportation system, while recognizing at the same time prospects for conversion of biomass to alternative synthetic forms of transportation fuels including gasoline and diesel in addition to the fuels required for aviation.

Two major hurdles—transport costs and unstable markets

These technologies had all been developed and tested extensively by the end of the first decade of this century, but the harnessing of biomass to decarbonization still faced two additional hurdles. The first is the high cost of its collection and storage at industrial sites. The second is the extreme volatility of oil and gas prices since the 1980s One comparison highlights the first problem: a tonne of anthracite coal weighs 850–870 kg. One cubic metre of baled and compressed rice or wheat straw weighs about 120 kg. The cubic metre of anthracite has an energy content of 6,000,000 kilocalories. The cubic metre of rice straw has an energy content of 420,000 kilocalories. Thus, a biofuels plant of any given capacity will need to spend 16 times as much on transportation of the fuel and provide 16 times as much storage space as would be required for coal. It is these costs that make a direct replacement of coal or other sources of fuels such as petroleum or natural gas with biomass difficult.

As is explained in chapters 16 and 17, in developing countries with large agricultural economies and an abundance of crop residues, this first hurdle can be crossed by using a two- stage gasification process, but it is a second hurdle that has so far proved insurmountable and has led to the virtual abandonment of the attempt to replace oil, coal, and natural gas with biomass as sources of transport fuels and industrial heat. This relates to the volatility and unpredictability of the price of crude oil and natural gas, and thus the ex-refinery prices of transport fuels, chemicals, and industrial hydrogen. In inflation-adjusted 2021 prices, after rising gradually from $18.94 a barrel in March 1946 to $24.35 in March 1973, the entry of OPEC into the oil market caused the price of crude oil to jump to $63.44 in February 1974, and to rise again to $141.45 in June 1980. Since 1980, it has swung up and down in roughly 6-year cycles, falling to $50.88 in January 1986, rising to a peak of $88.19 in September 1990, dropping to $30.08 in March 1999, rising again to $156.46 in April 2008, before crashing to $42.16 in February 2016 and rebounding, post COVID-19, to $116.16 in May 2022.

This volatility makes it virtually impossible for an investor to commit to biofuel production without some guarantee of stability in future earnings. Ethanol production took off in the US only after President George W. Bush's 2006 announcement of a subsidy of 51 cents a gallon on sales. A $38 billion subsidized credit programme to produce ethanol from sugar and molasses, announced by the Brazilian government in 2012, provided a further inducement. Without such backstopping, no private investor could responsibly consider making the large investment that biofuel production from crop residues and forestry wastes would require.

India's potential for generating biofuels and hydrogen from biomass

In India, there are four types of biomass residues that do not face the problem of the prohibitive transport and storage costs, since they are already being collected for other purposes. These are: (1) black liquor, a carbon-rich mix of chemicals that is the residue left from the bleaching of pulp in the production of paper; (2) rice husk and bran produced as products of rice milling; (3) bagasse, the finely crushed pulp left in sugar mills after all the sugar cane has been extracted, often called sugar cane trash; and (4) municipal solid waste. The agriculturally related industries currently use their biomass waste as a boiler fuel to meet their need for process heat. In all of them, however, replacing the boilers with gasifiers could provide not only the needed heat but also a synthesis gas that could be converted into a transport fuel of choice, or into green hydrogen.

Black liquor

The Indian paper industry produces more than 25 Mt of paper a year, in over 600 paper mills. Most of these are small and specialized, but 12 integrated modern mills that use the Kraft process for chemical processing to produce pulp account for the bulk of the output. Today, all of these recover the lignin removed from the pulp as a black liquor fluid. This contains 15–20% of carbon-rich solids, accounting for about half of the energy of the wood chips and pulp fed into the mill. The present method of recovering this energy is to dry the liquor until the solids make up 75–80% of the weight of the composite and then to burn the product in a recovery boiler to generate steam. A gasification system, however, could feed the black liquor directly into a high-pressure gasifier to produce syngas that could be used to produce electricity, steam, or any other gasification product, including biofuels and green hydrogen.[10]

The IEA convened a workshop in Ottawa, Canada in 2002 to consider the future prospects for biomass energy. The workshop reported that the global manufacture of paper and board in 2002 amounted to 91 Mt and was responsible for production of 170 Mt of black liquor measured as dry solids. This was sufficient to produce 48 Mt of methanol, via the black liquor gasification process.[11] By this yardstick, the Indian paper industry is capable of an annual production of as much as 13 Mt of methanol.

Rice husk and bran

Rice husk and bran make up 28–32% of the paddy that farmers harvest. In 2020–21, India produced 122 Mt of polished rice, slightly more than the average output in recent years. The yield of paddy has therefore been around 170 Mt, with rice husk and bran accounting for about 50 Mt. Today, the paddy is milled by an estimated 130,000 rice mills that separate 2–4 tonnes of rice bran and husk from the paddy per hour, extract 1 Mt per year of very high value rice bran oil, and burn the rest in their boilers. Here too, gasifying, instead of burning the rice husk and bran, could be used to produce not only an important source of heat but also a significant yield of synthesis gas that could be turned into a biofuel or hydrogen. The balance would be biochar which could be put to a variety of uses. Biochar from rice bran is of especially high potential value since 15–20% by weight would be represented by silica of high purity which could be sold as a raw material to PV panel manufacturers. The gasification of rice straw in a cyclone gasifier was a joint project of the KTH Royal Institute of Technology in Stockholm, Sweden, and the Indian Institute of Science in Bangalore (now Bengaluru) in 2009–10. The process was proved at the laboratory stage of testing but failed to progress to commercialization. Had this occurred the process could have been exploited to produce on an annual basis as much as 5–6 Mt of high-grade amorphous silicon.

Bagasse and sugar cane residues

The most abundant potential source of biofuels and green hydrogen, however, relates to the residues produced by the sugar industry. These include not only the bagasse left after crushing the cane to extract its juice, but also the sugar cane leaves and tops and residues left in the fields

Figure 12.1 Sugar cane 'trash' left behind in fields in the Philippine Islands.
Source: Hannes Grobe, 2006-04-07. https://commons.wikimedia.org/wiki/File:Sugar_cane_madeira_hg.jpg.

that are collectively described as sugar cane 'trash'. A picture of the 'trash' left in the field in a cane farm in the Philippines is shown in Figure 12.1. In India, the sugar cane stalk makes up 58% of the dry matter delivered to the sugar cane plant, leaves and tops 40%, and roots 2%.[12] The processing of 100 tonnes of sugar cane in a factory yields 10–12 tonnes of sugar, 30–34 tonnes of bagasse, 4–4.5 tonnes of molasses, converted normally into ethanol for the liquor industry, and 3–3.5 tonnes of press mud.[13] The bagasse is associated with up to 40% moisture and has to be almost completely dried before being fed into the boilers. The leaves and tops and other 'sugar cane trash' that have to be separated from the cane before crushing add up to another tonne or more of crop residues for every tonne of cane that is crushed.

Today, sugar mills burn the bagasse in modern boilers to produce the heat needed to distil the cane juice, and in other downstream processes required to produce white sugar. But these boilers produce four times as much heat as the sugar distillation and bleaching processes require.

So, around half of the sugar mills have set up power plants to produce and sell their surplus electricity to the grid. The boilers, however, can burn only the crushed and dried bagasse, and they work therefore only as long as the sugar cane crushing season lasts, typically 6–7 months of the year. But if sugar mills can switch from boilers to gasifiers to produce the heat needed for distillation, they will be able to gasify the remainder of the sugar cane waste as well, work throughout the year, and produce biofuels or green hydrogen the rest of the time. This will greatly increase their revenues in addition to generating a large amount of full-time employment. As much as 68% of the press mud created in the processing of the sugar is pure carbon. At present, there is no use for this product. Mill owners in India have tried to compost it and sell it as fertilizer to farmers and gardeners but have been defeated by the high cost of transporting the compost over more than a short distance.[14] Not only will adding the press mud to the gasifier fuel further raise its energy content, but also its sticky consistency makes it an ideal binder for use as a pelletizing agent.

India is the world's second largest producer of sugar cane. The annual output has increased from 350 Mt in the first decade of this century to 419.25 Mt in 2020–21. So, there is an equal quantity of sugar cane waste available for conversion into transport fuels or green hydrogen. The surplus electricity now being sold to the grid will be more than sufficient to meet the needs of the electrolysis and fuel-synthesizing plants. As the experiments carried out by Specht et al.[15] have shown, using an electrolyser to obtain the necessary quantity of oxygen and feeding the surplus hydrogen back into the reaction serves to double the yield of methanol from 0.3 to 0.6 tonnes from every tonne of biomass feedstock. Using a larger electrolyser to add hydrogen sufficient to convert all the CO_2 in the syngas will nearly double the methanol yield again, from 0.6 tonnes to 1.1 tonnes from every tonne of dry-weight biomass.

To sum up, under optimal conditions, India's sugar industry can additionally produce either 400 Mt of methanol-equivalent biofuels or an equivalent amount of green hydrogen, or a combination of the two. So, this one industry is capable of meeting a significant fraction of India's projected demand for energy in its transport sector in 2050, estimated at 450 Mt of fossil energy equivalent, using a combination of waste products derived from its manufacture of sugar.

Municipal solid waste

Municipal solid waste (MSW) is the bane of rapidly urbanizing developing countries. The World Bank estimated that the world generated 2.0 billion tonnes of MSW in 2016, forecasting that this would rise to 2.6 billion tonnes by 2030, climbing to 3.40 billion tonnes by 2050. Since most of the urbanization is taking place in middle- and low-income countries, the bulk of the increase over the next 30 years is likely to arise in these environments. The World Bank concluded that by 2050 the daily generation of waste in developing countries can be expected to increase by another 40%. But since populations in these countries are also increasing, the total amount of waste generated in these environments is projected to rise to an even larger extent: by a factor of 3 in sub-Saharan Africa while more than doubling in South Asia and the Middle East.[16]

MSW is the most variegated biomass source in the world. It incorporates up to 30% inorganic material and everything from plastic at the high end of the energy scale to kitchen waste at the low end. Its moisture content varies significantly from season to season. So, it is easily the most difficult form of biomass to gasify. But, somewhat paradoxically, it is the first biomass waste to have been gasified. This is because MSW is the one biomass that every city authority in the world is compelled to collect and to dispose of safely. In the 1990s, biomass gasification attracted the attention of industrialized countries because it completely eliminated a number of dangerous pollutants in the waste, such as dioxins and furans, that could not be eliminated by incineration. MSW gasification also attracted investors because no collection costs were involved and city authorities paid handsome 'tipping fees' for its disposal. In 2009, there was a large number of MSW gasification projects at various stages of planning and implementation in the US and Europe. And then the world went into recession. Suddenly strapped for money, city authorities one after another abandoned these projects and stayed with their extant incineration and landfill schemes. Today, the need to move out of fossil fuels into biofuels and green hydrogen has revived interest in MSW gasification all over the world, as indicated in our earlier discussion of the activities involving Fulcrum BioEnergy.

The need for waste disposal is most urgent in the low- and middle-income countries. In these, municipal authorities collect 48% of the waste generated in the cities and 26% of that produced in the surrounding semi-urban areas. But of the 48%, three-quarters is dumped on open patches of land a few kilometres outside the cities and towns. This is in stark contrast with the practice in high-income countries, where the corresponding figures are 96% for collection and 2% for open dumping.[17] According to the World Bank, South Asia generated 334 Mt of waste per day in 2016. Of this, India accounted for close to 80%, 260–270 Mt, 57% of which was vegetable matter with a high moisture content, rising to as much as 60% during some seasons of the year. Wood, paper, and plastic, with high energy content and almost no moisture, accounted for 21% of the waste. Inorganic glass, metal, and debris, with next to no energy content, was responsible for the balance. Overall, the energy content of the collected garbage per unit mass is probably around half that of wood. Based on this estimate and the results of the simulations by Specht et al.[18], gasifying MSW in India using an electrolysis system to deliver the required oxygen could be expected to yield 40–50 Mt of methanol or DME if no hydrogen is added and 130–150 Mt if sufficient hydrogen is supplied to convert all the CO_2 in the syngas. As indicated earlier, DME would be a perfect replacement for LPG gas for cooking in urban India.[19]

The benefits of biomass gasification in India would be significant in terms of a reduction in the emission of greenhouse gases, but these benefits would be dwarfed by the improvements in health, water, and air quality that would follow. A recent simulation in Sweden of a black liquor gasification plant using carbon capture to eliminate residual CO_2 emissions indicated that 1 dry tonne of the organic fraction of MSW could generate 72 kg of hydrogen and take out over 700 kg of CO_2 from the atmosphere.[20] Based on this finding, if India's MSW production doubles from the present 270 Mt to 540–550 tonnes by 2050 and its composition (57% kitchen waste with average 50% moisture content, 21% wood and plastic, and 22% inorganics) remains unchanged, the dry weight of the future organic fraction of the MSW will be around 267 Mt. This would suffice to produce 294–300 Mt of methanol-equivalent biofuels or 190–195 Mt of green hydrogen.

We conclude, therefore, that India's demand for carbon-free energy in 2050 could be realized comprehensively using a combination of the renewable sources of energy from wind and sun discussed in Chapter 10, the green hydrogen described in Chapter 8, and the sources of biomass discussed here.

Notes and References

1. https://www.whichcar.com.au/news/how-many-cars-are-there-in-the-world (viewed 7/9/2022).
2. Masson-Delmotte, Valérie, et al. (2018). Global warming of 1.5 °C. An IPCC Special Report on the impacts of global warming of 1.5 °C. https://www.ipcc.ch/site/assets/uploads/sites/2/2022/06/SR15_Full_Report_HR.pdf (viewed 7/9/2022).
3. NITI Aayog (2017). Energizing India. https://niti.gov.in/sites/default/files/energy/Energising-India.pdf (viewed 7/9/2022).
4. United States 2021 Aviation Climate Action Plan. https://www.faa.gov/sites/faa.gov/files/2021-11/Aviation_Climate_Action_Plan.pdf (viewed 9/9/2022).
5. Quaak, P., Knoef, H., and Stassen, H.E. (2010). Energy from biomass: a review of combustion and gasification technologies. The World Bank.
6. Specht, M., et al. (1999). Synthesis of methanol from biomass/CO_2 resources. In: Proceedings of Conference on Greenhouse Gas Control Technologies.
7. Ibid.
8. World's first commercial-scale household waste-to-fuels plant completed. Big Report 2022. https://www.energylivenews.com/2021/07/08/worlds-first-commercial-scale-household-waste-to-fuels-plant-completed/ (viewed 9/9/2022).
9. International Energy Agency: Biofuels—What is the role of Biofuels in clean energy transitions? https://www.iea.org/energy-system/low-emission-fuels/biofuels
10. National Energy Technology Laboratory: Black Liquor Gasification. https://netl.doe.gov/research/Coal/energy-systems/gasification/gasifipedia/blackliquor (viewed 13/9/2022).
11. IEA Bioenergy: Black Liquor Gasification: Summary and Conclusions from the IEA Bioenergy ExCo54 Workshop. https://www.ieabioenergy.com/wp-content/uploads/2013/10/Black-Liquor-Gasification-summary-and-conclusions1.pdf (viewed 13/9/2022).
12. Kumar Kanaujia, Anoop and Mohan, Narendra (2021). Bio-Energy from Indian Sugar Industry: A Sustainable Renewable Energy Future. *International Journal of Engineering Research & Technology* 10(05).
13. Singh, P. (2020). Sugar Industry: A Hub of Useful Bio-Based Chemicals. In: Mohan, N. and Singh, P. (eds) *Sugar and Sugar Derivatives: Changing Consumer Preferences.* Springer. https://doi.org/10.1007/978-981-15-6663-9_11 (viewed 15/9/2022).
14. Information provided to the authors by the Indian Sugar Mills Association, 2011.
15. Specht, M., et al. (1999). Synthesis of methanol from biomass/CO_2 resources. In: Proceedings of Conference on Greenhouse Gas Control Technologies.
16. Silpa, Kaza, et al. (2018). *What a Waste 2.0: A Global Snapshot of Solid Waste Management to 2050.* World Bank, pp. 69–75. https://openknowledge.worldbank.org/handle/10986/30317 (viewed 15/9/2022).

17. Ibid.
18. Specht, M., et al. (1999). Synthesis of methanol from biomass/CO_2 resources. In: Proceedings of Conference on Greenhouse Gas Control Technologies.
19. Shweta Miriam Koshy: Overcoming India's clean cooking challenge. *Down To Earth*, 26 December 2019. https://www.downtoearth.org.in/blog/energy/overcoming-india-s-clean-cooking-challenge-68562#:~:text = The%20annual%20usage%20varies%20from,hour%20(kWh)%20of%20electricity (viewed 20/9/2022).
20. Wang, Shule, et al. (2022). Renewable hydrogen production from the organic fraction of municipal solid waste through a novel carbon-negative process concept. *Energy* 252:124056.

13

Correlating Decarbonization with Population Density and Economic Need

In the preceding chapters we presented, in some detail, the potential for India to shift its energy base from fossil to renewable sources, specifically wind, solar, and biomass, without jeopardizing its future economic growth. More specifically, we showed that this shift could not only provide many times more than the country's current demand for energy, but also could accommodate in addition multiples of the demand projected for 2050. But these are estimates of *potential*. They tell us what India could achieve. They fail, however, to identify the policies future Indian governments will need to follow in order to realize this potential.

This chapter and the chapters that follow are devoted to outlining the policy measures that will be required if the country is to realize its goals for a sustainable future. The first step in identifying these measures is to rediscover and acknowledge the importance of land. Acquiring and taming land has been the first goal of just about every venture that human beings have embarked on since the end of the last Ice Age. So it is not surprising that the great battles of antiquity occurred over the control of land. But land began to lose its salience in the 18th century when fossil fuels were first harnessed for industrial production, heating, and transport. Coal, mined in only a handful of places, could be shipped hundreds of kilometres to just a few industrial centres that could meet an entire country's needs. Oil and gas could be piped or shipped efficiently for even longer distances. Therefore, from being the prerequisite for growth and prosperity, land became secondary to the development process, required only to provide food and recreation for an ever-expanding population.

Sun, Wind, and Biomass. Prem Shankar Jha and Michael B. McElroy, Oxford University Press.
 DOI: 10.1093/9780198944768.003.0013

Today, with the end of the fossil fuel era in sight and the world turning increasingly to solar, wind, and biomass energy, land has once more regained its primal importance. But this is happening in a world that is almost unimaginably different from the one that existed at the dawn of the Industrial Revolution. Estimates of the global human population in 1750 range from 700 million to 814 million. In 2022 it crossed 7.75 billion.[1] There are therefore ten persons on the planet today for every one who lived in 1750. And, except in the first generation of industrialized countries, the majority still live on the land. So how much land can the world spare for generating energy?

For vacant land the obvious place to look first is where it is least inhabited today, i.e. in the desert. Desert surfaces are exposed to a little more than twice the solar radiation received by humid regions and lose almost twice as much of their energy budget in the form of heat at night. While their mean annual temperatures range from 20 to 25 °C, the mean diurnal change can vary from a maximum of 43.5–49 °C to a minimum of –18 °C.[2] Not surprisingly, the first solar-powered electricity plants were set up in the Mojave Desert of the US in the 1980s.

Even a cursory estimate of the desert area of the world shows that the land available for generating solar power is sufficient to meet the present and future needs of the world several times over. In 1977, UNESCO introduced a simple definition of 'desert'.[3] This was any area in which the annual precipitation of rain or snow is less than the annual evapotranspiration (a portmanteau term that includes both evaporation of water into the atmosphere and transpiration of water through plants). Defined in this way, 170 million km^2, a third of the 510 million km^2 of the Earth's land surface, can be classified as desert.

Excluding 28 million km^2 in the Arctic and Antarctic that are not capable of sustaining human life, there are 142 million km^2 of habitable 'hot' and 'cold' deserts on the planet. A seventh of this, 20.8 million km^2 in all, is located in 23 large desert environments that occupy 50,000 km^2 of the Earth's surface. These include the Thar Desert that straddles the India–Pakistan border. The remaining 121 million km^2 are located in smaller arid and semi-arid 'desert areas' of all sizes and shapes across the rest of world. There is hardly a single country, outside of northern Europe, that does not have a sizable amount of desert land. Even the island of Oahu in Hawaii has a significant mini desert at its western end.

The UNESCO definition of desert remains the one most commonly used even today. But in terms of framing a strategy to combat climate change it is somewhat misleading. For it includes anomalies that it failed to rationalize even from the day of its original introduction. The most important is that many of these supposed deserts are not only inhabited but also support cultivation because they receive their entire year's quota of snow or rain within a short 'growing season' in which precipitation is significantly greater than evapotranspiration. This surplus moisture is sufficient to sustain cultivation or to provide forage for cattle for most if not all of the year.

A second reason is that it completely ignores geological anomalies, such as the Okavango Delta in the Kalahari Desert, which represents the last vestiges of a much younger Planet Earth and is therefore, literally, a treasure of inestimable value. The Okavango Delta is kept alive by a river that rises in the mountains to its west, a river that has continued to pour 1,100 trillion litres of water a year into a 'sea' that dried up a hundred million years ago. This has created what may be the single most perfectly preserved flood-land habitat left in the world.

So, after prolonged research, in 1993 the FAO added a second criterion for the definition of desert: the length of the growing period (LGP), a concept the FAO had developed earlier for use in its study of Agro-Ecological Zones. In this formulation, the growing period starts when rainfall exceeds half of the daily potential evapotranspiration. Allowing for short breaks, the LGP ends on the day after rainfall falls below half of the daily potential plus the time required for the moisture level in the soil to fall by another 10 cm, i.e. until it is beyond the reach of crop roots. The FAO then subdivided this definition into regions with: (1) a growing period of less than 1 day, which it called *hyper-arid* (i.e. 'true' desert); (2) regions with less than 75 such days, described as *arid*; (3) environments with growing periods of from 75 to less than 120 days, identified as *dry semi-arid*; and (4) regions with growing periods of 120–180 days, classified as *moist semi-arid*.

A third even simpler definition, which tallies fairly closely with that of the FAO, is based on rainfall alone: it defines hyper-arid 'true deserts' as areas that receive less than 25 mm (1 inch) of rainfall a year; arid regions as those that receive between 25 and 200 mm (1 and 8 inches) a year, and semi-arid regions as those that receive between 200 and 500 mm

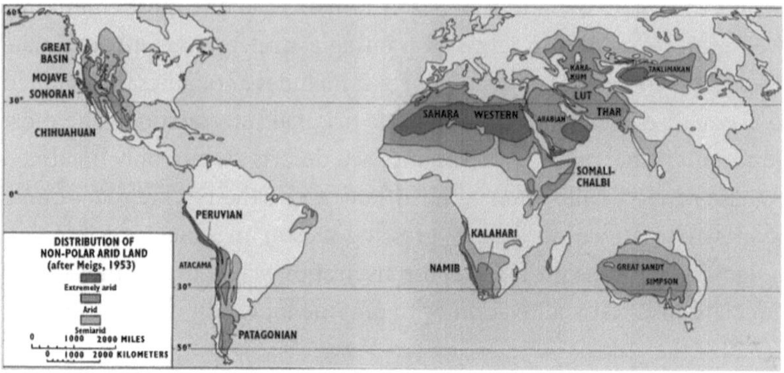

Figure 13.1 Distribution of non-polar arid land (after Meigs, 1953).
Note: Not to scale; for representational purposes only.
Source: US Geological Survey. https://pubs.usgs.gov/gip/deserts/what/world.html.

(8 and 20 inches) of rain per year.[4] Both of these definitions include only the hyper-arid and arid categories in their definitions of 'desert'. So they bring the estimated desert area of the planet down to about a fifth, as against a third, of the Earth's land surface. But even this is a vast area that covers more than 100 million km^2. The remaining 70 million km^2 falls in the *moist semi-arid* zone which receives rainfall from 120 to 180 days in a year, but still loses more moisture through evapotranspiration than it receives through seasonal snow or rainfall. Most of this land is to be found in the low-income developing countries, between 30 degrees north and south of the Equator. But almost none of it is uninhabited. This is the zone in which the major part of India's land surface is located. So, bringing it under solar or wind power development will proceed most rapidly if the present owners and users (if they are nomads) are made co-sharers in the benefits that the new power generation schemes will bring. Locations of the major desert areas of the Earth are indicated in Figure 13.1.

How much of India's renewable energy potential can be realized?

India is one of the two or three most densely populated countries in the world. But the availability of land was not a significant constraint on development so long as coal, oil, and gas were the primary sources of commercial energy. As Table 13.1 shows, the land area needed for setting up a coal-fired power plant, including facilities required for management and maintenance, can range from 670 hectares, i.e. 6.7 km^2, for a 1,000-MW coal-fired power plant to 1,150 hectares, i.e. 11.5 km^2, for a 4,000-MW plant. This amounts to accommodating 149–347 MW of generating capacity per km^2. Against this, utility-scale SPV plants in India have required around 4 hectares per MW, after accounting for the support facilities needed for maintenance and administration. One kilometre squared of land has therefore accommodated only 25–30 MW of SPV power-generating capacity. The figure for wind power is even lower, at barely 4 MW of generating capacity per km^2. In terms of delivered power, the disparity is even greater. For, while coal-based plants have worked for decades at 70–75% of their nameplate capacity, SPV plants averaged only 17–20% in 2019 and wind farms 20–22%. In sum, SPV plants need not 5 times but up to 20 times as much land to deliver the same amount of power as a coal-fired plant with the same nameplate generating capacity. Wind power generation requires even more land, but, without needing to displace more than a fraction of

Table 13.1 Land requirement for various equipment/system facilities in coal-fired power plants (in hectares).

Equipment/ system facilities	**2×500 MW (40% ash)**	**3×660 MW (40% ash)**	**5×660 MW (40% ash)**	**6×660 MW (40% ash)**	**4×800 MW (34% ash)**	**5×800 MW (34% ash)**
Main power plant						
i) Power plant	450	640	820	940	790	880
ii) Green belt for power plant	150	210	270	310	260	290
Total for main plant	600	850	1090	1250	1070	1170
Ash dyke including green belt	500	855	1370	1630	1000	1200
Other facilities outside plant area	220	245	250	250	250	250
Township	100	100	150	150	150	150
Grand total	1420	2050	2860	3280	2470	2770

Source: Report of the Land Requirement of Thermal Power Stations: Chapter 6 Central Electricity Authority New Delhi, December 2007.

land from its present use. So, as India begins its transition out of fossil fuels in earnest, land is once again becoming a crucially important factor for both energy generation production and economic development.

The availability of land

India has a geographical area of 328 mha (3.28 million km^2). Of this, the usable area amounts to 308 mha (3.08 million km^2). The balance of about 20 mha (200,000 km^2) consists mostly of waterbodies and marshland. Of the usable area of 308 mha, 68 mha is forest land, much of it denuded but controlled by state forest departments, an environment that no one would dare touch for any other purpose. Another 23 mha is urban land and land reserved for non-agricultural uses.[5] Of the remaining 217 mha, 142 mha is cropland. The final 75 mha consists of wasteland (55.7 mha), grasslands, and fallow land.[6] Apart from some of the wastelands, nearly all of this land is privately owned. So the availability of land for wind and solar power plants will depend mainly on how much of the land the owners are willing to divert from current use to allow for development of power generation alternatives.

Over the past two decades there have been several studies of India's solar and wind power generation capacity based on the quality of its physical resources and the availability of land. But most of these studies have been hampered by the lack of detailed day by day and hour by hour data on insolation and wind flows over its 3 million-plus km^2 land surface. But, as has been described in Chapters 9 and 11, this knowledge barrier was broken in a study by Song et al.[7] at the School of Engineering and Applied Sciences of Harvard University. This study used a compilation of hourly data obtained from analysis of historical meteorological data.[8] They concluded that the annual potential of offshore wind, onshore wind, and SPV for the generation of electricity in India amounted to 1,546 TWh, 22,200 TWh, and 20,900 TWh respectively. Taken together, this is almost 30 times the consumption of electricity in 2018–19.[9] Results for the potential capacity factors for wind and solar derived from this study are presented in Figure 13.2.

As was pointed out in Chapter 9, India is fortunate to have a high degree of congruence between the areas with the maximum solar and

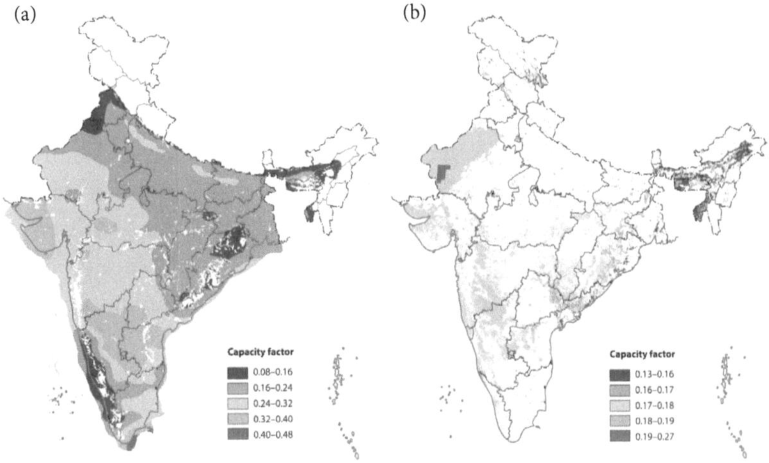

Figure 13.2 Potential capacity factors for generation of power from wind and solar systems.

Note: Not to scale; for representational purposes only.

Source: Song et al., Deep Decarbonization of the Indian Economy: 2050 Prospects for Wind, Solar, and Green Hydrogen. *iScience*, 25 (2022).

wind generation potential. This is most evident in the southern and western parts of the country, where large portions fall into the moist semi-arid category of land described above. Its exploitation for solar energy generation is constrained, however, by the density of settlement on the land and the precariousness of the livelihoods it generates in its current uses—livelihoods the current residents can ill afford to lose.

In the semi-arid states of southern and western India, the population density ranges from 555 per km^2 in the relatively well-watered state of Tamil Nadu to 236 per km^2 in Madhya Pradesh, accountable for as much as 200 per km^2 even in the desert state of Rajasthan. The average for the eight states of this region is around 300 per km^2 and, as Chapter 14 elaborates, their populations are mostly rural and the lives of their residents are precarious, with high rates of out-migration to cities and high levels of suicide. The owners of the land will therefore need to be persuaded to put their land to an altogether different and novel use. For that, India will need to frame policies that will make the land owners gainers not losers from the introduction of the new technology.

Solar power potential

We have noted that the land required for SPV installations is greater than the land required for wind systems. The first, and indeed most obvious, place in which to locate a solar power plant would be on dry semi-arid land that has few other uses but has a minimal supply of water. Most of the 55.7 mha (557,665 km^2) of the land classified as wasteland, meets this criterion. But 31% of it, i.e. nearly 173,000 km^2, is located in the higher reaches of the Himalayas, in the states of Jammu and Kashmir, Uttarakhand, and Arunachal Pradesh in the north-east. Of these areas the most easily exploitable is a cold desert measuring 38,000 km^2 in Aksai Chin. But Aksai Chin has been in the eye of a dispute between China and India, that has so far ruled out its use for any but military purposes.

Much of the remaining 135,000 km^2 is well suited for producing solar power because of its high insolation and absence of rain and dust., but since transmission costs from these regions to the plains of India are prohibitively high, these can at best be used to generate electricity for local use. This is because the residents of these regions need electricity not only for lighting, study and work, but also for home heating during 6 to 7 months of the year. At present, that need is met by burning coal and increasingly scarce firewood in braziers or by heavily subsidized electricity when it is available, which it usually is not.

The balance of the land classified as wasteland, 384,665 km^2, is located in the eight central semi-arid regions of the country that have both high and stable levels of insolation together with a high availability of wind power.[10] The greatest single chunk is in the Indian portion of the Thar Desert, with 238,000 km^2 sited in Rajasthan. As the National Institute of Wind Energy pointed out in 2010, much of this area is presently too remote and poorly developed to make it possible to transport the giant blades, rotors, and turbines needed for wind power generation. SPV power plants do not suffer from this drawback. Solar panels come in easy-to-transport sizes, weigh relatively little, and require only light supporting structures. What is more, as the importance of wind and SPV power has come to be recognized, so has the need to build the roads, bridges, and power and water lines needed to tap the potential of these areas. By 2050 it should be possible to bring three-quarters of the 238,000 km^2 of the available wasteland in these states under solar 'farms'.

At 30 MW per km^2, this would facilitate the installation of 5,355 GW of generating capacity. At an average capacity factor of 19%, this is capable of generating nearly 8,912 TWh of electricity annually. That is over five times the consumption of electricity (1,547 TWh) in 2018–19, and almost equals the likely demand of 8,323 TWh estimated for 2047. Yet it is no more than and more than half of the wind power potential estimated by the Harvard study described above. A map of India's SPV potential is presented in Figure 13.3.

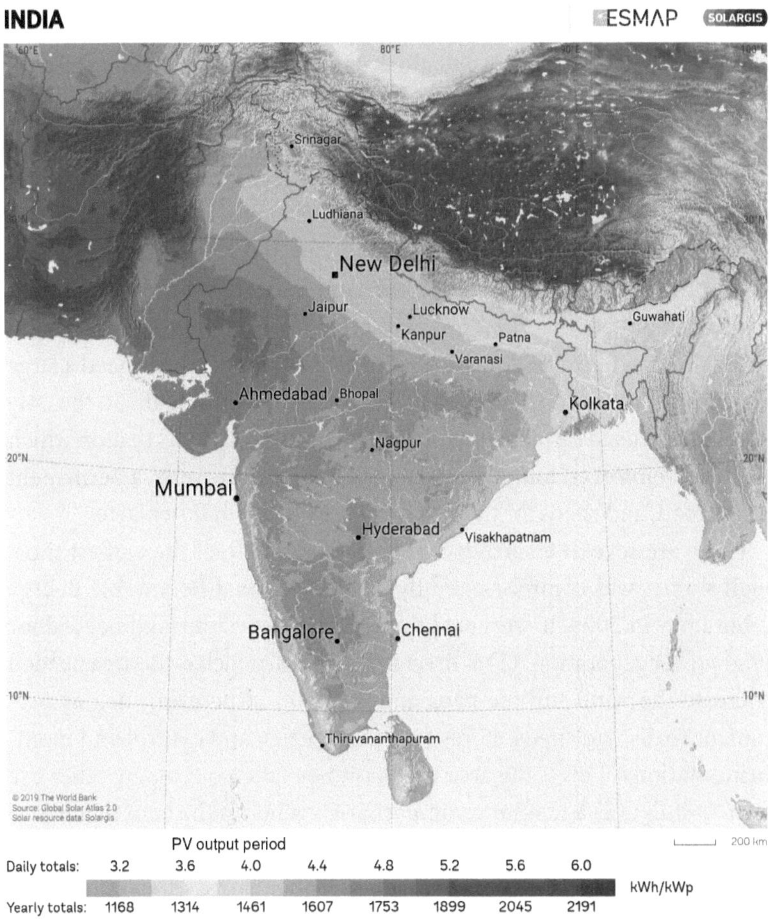

Figure 13.3 Photovoltaic power potential.

Source: From SolarGIS © 2013 Geo Model Solar, CC BY-SA 3.0. https://commons.wikimedia.org/w/index.php?curid = 33459572.

Wind power potential

As the Harvard study's map of wind power potential shows, outside the cold regions of the Himalayas, the potential for wind is almost entirely concentrated in a swathe of land that covers the western and southern half of the country. In this entire area, the study found that, except in the wind shadow of the Western Ghats, the capacity factors of turbines set up in the area ranges from 24 to 40%. But the exploitation of this enormous technical potential is constrained in practice by the density of settlement on the land, by the revenue the land generates in its current uses, and by the compatibility or incompatibility of some of these uses with either SPV, CSP, or wind power installations. Given the relative unsuitability of most of the wasteland in the country for wind, as opposed to solar, power generation and that all but about 20 mha of current and permanent fallow lands are under the plough, it would seem that the opportunity to realize this enormous potential is extremely limited. But a straightforward comparison with the land available for solar power generation is misleading because whereas all of the land needed by an SPV plant has to be taken out of other uses, only a fraction of the land needed by a wind turbine needs this to be done. This is because Wind turbines need a large open space around them to ensure an unimpeded flow of air. So the only land that needs to be taken out of its current use entirely is that on which the deep, reinforced foundation on which the turbine rests, a permanent access road, and sheds for the storage of spares and maintenance.

There are several estimates for these needs. One of the earliest thorough studies was commissioned by the US National Renewable Energy Laboratory in 2009. It estimated three properties of the land needed for wind power generation: (1) a direct impact area, which is the area needed to create the wind turbine pads and ancillary structures, such as permanent roads, toolsheds, stores for components, and control and monitoring stations. This is the area that has to be taken out of any other use permanently; (2) a temporary impact area, which is the land that has to be taken out of other uses while the wind turbine is being built. This consists of 'lay down' areas for the huge components and other equipment that has to be assembled at the site, the cranes, elevators, workshops, power generators, scaffolding, and temporary roads needed for assembly and erection.[11]

The rest of the area can go back to its original use, provided only that it is kept free of tall buildings and plantations of high-growing trees, to ensure the maximum, uninterrupted flow of wind to the turbines. A study of wind energy farms in the US by the US' National Renewable Energy Laboratories found that the direct impact area permanently withdrawn from other uses came, on average, to no more than 0.3 hectares i.e. 1.2 acres of land per turbine. But the total area that had to be kept free of tall structures permanently varied from 12 to 56 hectares, with an average of 33 hectares. This meant that no more than three wind turbines on an average could be installed on one km^2 of land.

This requirement is not as severe as it sounds, because even with a turbine mounted 80 m above the ground, there is a clear 30 m or more of space between the lowest rotation point of its blades and the ground. That is higher than an eight-storey building. As wind farm operators in the US have found, spreading the financial benefits from the turbine through the neighbourhood via a 'Good Neighbour' royalty payment to all the other farmers with holdings around the wind turbines is an effective way to ensure freedom in the future from litigation, not to mention blackmail.

A more recent estimate by US-based National Wind Watch Inc. concluded that to maximize the impact of the wind on the blades and avoid turbine–turbine interference, wind turbine towers needed to be placed 10 rotor diameters apart in the direction of the wind and 3 rotor diameters apart in every other direction. Placed in line perpendicular to the wind, the GE 1.5-MW model would need at least 12.8 hectares of clearance per turbine and the 3-MW Vestas V90 would need 31.4 hectares. However, in an array designed to take advantage of winds from more than one direction, the GE turbine would require 32.8 hectares and the Vestas turbine 44.4 hectares. These estimates are consistent with the recommendations of the NREL.[12]

In India, the vast majority of turbines receive their wind from at least two directions, as indicated in Figure 13.4. So, the higher estimates of 32.8 and 44.4 hectares are the relevant ones for estimating wind power potential in India. This means that no more than two Vestas V-90 turbines (or any other 3-MW turbine) can be installed on every kilometre squared of land. This gives a generation capacity of 6.6 MW per km^2 as against the 3.8 MW assumed in the Harvard study. In what follows, therefore, two estimates are discussed for the wind power potential in India.

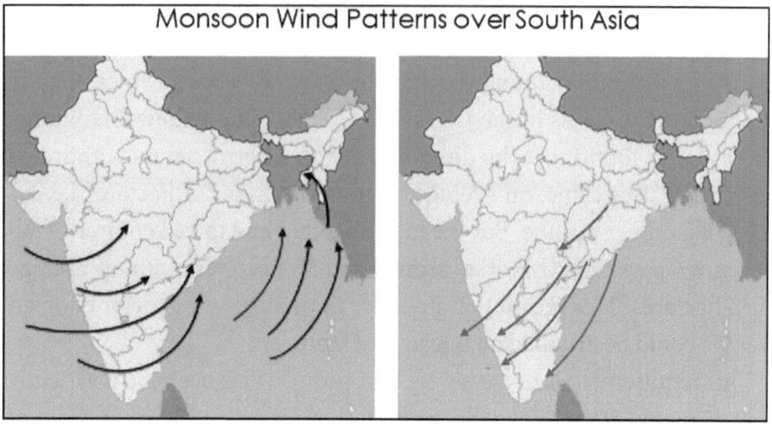

Figure 13.4 Prevailing wind patterns during summer and winter.

The potential for generating wind power is concentrated almost entirely in the 'moist' semi-arid zones of western and central India, notably in the states of Gujarat, Rajasthan, Madhya Pradesh, Maharashtra, Chhattisgarh, and Jharkhand and in the four southern states of Andhra, Telengana, Karnataka, and Tamil Nadu. The lower two-thirds of this region receive steady winds from both the south-west and the north-west monsoons, which blow for close to 8 months from June until the end of January. These states are therefore well suited for harnessing a combination of solar and wind power. The winds blow strongly between June and January at a time when cloud cover from the south-west and then the north-east monsoons reduces solar radiation to a minimum. The winds die away between February and the end of May when there is seldom a cloud in the sky.

As indicated in Table 13.2, the total cultivated area in these ten states amounts to 110.1 mha. Of this, 27 mha are irrigated and nearly all of this is double cropped. One crop at least is a paddy (rice) crop which requires the fields to be flooded for several weeks a year. As the flooding will soften the soil and weaken the foundations of the wind turbine, this land cannot be used for wind power generation. But 83 mha are dependent solely on the rains for their productive crops, and global warming has been making the rains progressively less reliable for more than three decades. Farmers in these rainfed agricultural areas are therefore in urgent need of an alternative, secure source of employment and income. Their desperation is reflected by the rising rate of suicides.

Table 13.2 Properties of farming in selected Indian states (in million hectares).

State	Total cropped area	Irrigated land	Dryland farming
Andhra and Telengana	16.6	4.9	11.7
Karnataka	12.2	2.4	9.8
Tamil Nadu	6.5	2.7	3.8
Gujarat	9.9	3.2	6.7
Rajasthan	21.1	5.1	16.0
Madhya Pradesh	15.8	4.2	11.6
Maharashtra	19.8	3.4	16.4
Chhattisgarh	5.1	0.85	4.25
Jharkhand	3.2	0.25	2.95
Total	110.2	27.0	83.2

Wind turbines can provide just the lifeline that the residents of these states require to stabilize their incomes and their lives and bring down the suicide rate. The crisis that farmers are facing in both the dryland farming and irrigated areas of the country is described in Chapters 14 and 15. Here it is sufficient to say that, as has happened in Gujarat to the owners of land that has been acquired or leased for setting up SPV utilities, the 'Good Neighbour' payment policy followed in the US is unlikely to work in India because of the tardiness of legal processes and the expense of litigation. Indian villagers learned this lesson at their cost in the first decades after Indian independence when state governments acquired vast amounts of land to build irrigation and hydroelectric power systems and reneged on promises of compensation, leaving hundreds of thousands of their citizens penniless. Since then, they have learned to stall such projects, often for years on end, by going to the courts appealing against the acquisition orders. If the governments of these states learn from their experience with hydroelectric projects and make royalty payments to all farmers in the coverage area of wind turbines as a part of the electricity tariff, there is no reason why all of these 83 mha should not come under wind power generation. Assuming that 10% of these 83 mha is withheld for investments in wind farms and that the land area that needs to be kept free is 0.68 km^2 (68 hectare) per turbine as estimated in the Harvard study, the remaining 75 mha, i.e. 750,000 km^2 of farmland will be able to generate 8,100 TWh of electricity, with 2.5-MW turbines working at an average capacity factor of 32%. However, if the wind turbines are upgraded to 3 MW as with the Vestas 90 and placed within a coverage area of 0.45 km^2 (45 hectares) the installed generating capacity will rise to 6.6 MW/km^2. With an average capacity factor of 32%, the power generated from wind would rise in this case to 14,454 TWh.

This is not all of the land that can come under development for wind turbines. There are also 37 mha of temporary and permanently fallow land, most of it adjoining the dryland farming areas, of which at least three-quarters could be brought under wind farm development with inducements similar to those offered to dryland farmers. This would add another 1,831 GW to the generation capacity. At a capacity factor of 32% this would be capable of generating 5,134 TWh of power annually. This would raise the generation potential from the eight southern and western states to 19,188 TWh at the higher capacity factor of 32% and 15,303 TWh at the lower capacity factor of 25%. These on-the-ground assessments of solar and wind power generation capacity in India are consistent with the estimates from the Harvard study.

Notes and References

1. https://statisticstimes.com/demographics/world-population.php.
2. University of California Museum of Paleontology, Berkeley: The Desert Biome. https://ucmp.berkeley.edu/exhibits/biomes/deserts.php (viewed 20/9/2022).
3. UNESCO's MAB Technical Note 7 of 1977, with an accompanying map.
4. Laity, Julie J. (2009). *Deserts and Desert Environments.* Volume 3 of Environmental Systems and Global Change Series. John Wiley & Sons, pp. 2–7, 49 (viewed 20/9/2022).
5. Land use statistics at a glance: 2008–09 to 2017–18. Directorate of Economics and Statistics, Department of Agriculture & Farmers Welfare, Ministry of Agriculture & Farmers Welfare, Government of India, July 2021. https://desagri.gov.in/document-report-category/land-use-statistics-at-a-glance/.
6. Land Use in India: wasteland amounting to 557,65 km^2, of which 1,191,000 km^2 is desert and other uncultivable land; 232,220 km^2 is abandoned land and permanent fallow land; and 139,400 km^2 is other unclassified wasteland. Land sequestered from agriculture amounts to 912,500 km^2. This consists of 688,000 km^2

of forest land; built-up areas, roads and railways, and industrial land; and 1,564, 630 km^2 of agricultural land, of which 661,030 km^2 is irrigated, 903,600 is under dryland farming, and 17,800 km^2 is under organic farming.

7. Song, S., et al. (2022). Deep decarbonization of the Indian economy: 2050 prospects for wind, solar, and green hydrogen. *iScience* 25(6).
8. Ibid.
9. The total consumption of electricity generated by state-owned and private producers was 1,547 TWh.
10. National Wind Watch: FAQ—Size. How much area is required for a wind power facility? https://www.wind-watch.org/faq-size-p.php (viewed 21/9/2022).
11. Land Needed for Wind Power Plants in the US. https://www.nrel.gov/docs/fv09osti/45834.pdf (viewed 23/9/2022).
12. Ibid.

14

Combining Decarbonization and Economic Growth to Open a New Future for the Rural Poor:

1. What Solar and Wind Energy can do:

In Chapter 3 we described how the stalling of industrial growth after 2011 and the rising pressure of unemployment have sown a fear of the future and a growing desperation in the youth that is eroding the stability of Indian democracy. As the McKinsey 2020 report cited in that chapter underlined, the only antidote that policy makers have been able to think of is to bring back the rapid growth that India had experienced for nearly two decades, from 1993 to 2012. But even a return to the very high growth of this period is not likely to solve the problem of unemployment because of the profound changes in technology that have taken place over the past several decades, notably the automation of manufacture and the computerization of middle-level jobs in the services sector. These have made many categories of blue and white collar workers redundant in India, as elsewhere, and thereby steadily deepened the cleavage between product and employment growth.

The political and social tensions that this cleavage has created are still manageable in the industrialized countries, where overt unemployment is limited, populations have virtually ceased to grow, and social security systems created a century ago have not withered away. In them, annual GDP growth has averaged 2.3% since 2013, while the related consumption of primary energy per unit of GDP has declined by 2%. In addition, the carbon intensity of the energy they are consuming is declining by 0.3% per annum. The OECD countries have therefore already crossed their carbon emission peaks and only need to eliminate the carbon

Sun, Wind, and Biomass. Prem Shankar Jha and Michael B. McElroy, Oxford University Press.
© Prem Shankar Jha and Michael B. McElroy 2025. DOI: 10.1093/9780198944768.003.0014

content of their existing energy consumption. But these tensions are far less manageable in the low- and middle-income countries. Therefore, while decarbonization technologies that also create employment and income growth are sure to receive wholehearted support from their people, those that aim only at decarbonization will find far fewer takers.

So far, all of the technologies that have found widespread acceptance in the industrialized countries—solar and wind power generation, battery storage, the creation of 'smart' grids, the production of green hydrogen via electrolysis, and electricity-driven transport vehicles—have been highly capital intensive. In their rush to develop new technologies for reducing their own carbon emissions, the industrialized countries have paid little attention to their suitability for developing countries. Two examples are the rush to develop 'carbon capture and storage' for 'clean coal' power plants that developed in the 2000s and a similar rush that is developing now to produce hydrogen to meet the heat requirements of industry. Both of these require sizable investment but yield no increase in output. In fact, in terms of consumable goods and services, they do the opposite. So it is no surprise that they have made barely a dent in the investment plans of even the most capital-rich nations, let alone the developing countries.

The developing countries have no option therefore but to find ways to harness on their own the technologies that can combine decarbonization with economic growth. The Third Working Group of the IPCC's Sixth Assessment Report said as much when it advised the world that 'strategies devised in the context of sustainable development, equity, and poverty eradication, and . . . the development aspirations of the societies within which they take place, [would] be more acceptable, durable and effective'.[1] But the COP 27 conference at Sharm el-Sheikh showed that the warning had fallen on deaf ears, for its Implementation Report estimated that the world would need to invest $4–6 *trillion* dollars a year for the next 30 years to attain net zero emissions by 2050,[2] and that the developing countries would need to spend close to $750 billion *a year* for the next 8 years to meet even the limited targets for decarbonization that they had set themselves after the Paris Summit.[3] But it did not have a word to say about how this money needed to be spent.

Against these huge requirements, the world had been spending only $803 billion a year on harnessing renewable energy technologies. It is therefore not surprising that shortly after COP 27 ended, Hans-Otto Portner, the Chairman of the IPCC, warned the world that without a drastic change of strategy it would cross the point of no return—of 1.5 °C of global warming by 2030 instead of by the end of the century.[4]

In India, Portner's warning fell on deaf ears. Within days of the end of COP 27, the government flooded the newspapers with rosy predictions that by 2030 India would be generating 62% of the electricity it consumes from non-fossil sources, including hydro and nuclear power. Installed power-generating capacity would rise from the current 409 GW to 820 GW. Of this, 50% would come from solar, wind, hydro, and nuclear power.

Solar- and wind-generating capacity would increase from 100 GW in 2022 to 500 GW in 2030. Its spokespersons failed to mention that 100 GW was a little more than half of the target of 175 GW (including 100 for solar and 60 for wind) that Prime Minister Modi had set for 2022 shortly after he came to power in 2014. If the government made no attempt to analyse the causes of its failure, there was little or no likelihood that it would meet its 2030 target either.

In this and Chapter 15 we discuss how wind, solar, and biomass could enable India and most other developing countries to end their dependence on fossil fuels and at the same time achieve their developmental goals. This chapter describes the policy changes that are required to do this with wind and solar energy. The next chapter addresses the potential for biomass.

The contribution that wind turbines can make to economic growth and employment

Wind power was the first form of renewable energy that the Indian government sought to harness. But it did so gingerly, with an implicit belief that this would be possible mainly on uncultivated land because cultivators would not put up with the disruption of agriculture and animal husbandry that it would involve. That is one reason why, despite having itself commissioned the Wind Force study described in chapter 8, the current official estimate of the onshore wind power potential of the country—695 GW—continues to be based on the National Institute for Wind Energy's assumption that only 2% of India's land surface will be available for setting up wind turbines. To place this number in context, we would note, as discussed in Chapter 9, that 817 GW of onshore wind power potential is estimated to be available in Rajasthan alone, with additional resources of 499 GW and 460 GW available in the neighbouring states of Madhya Pradesh and Maharashtra respectively.

However, in Gujarat and Tamil Nadu, where wind farms were first established, it has brought significant benefits to the rural community. In a perceptive article on the impact of wind turbines in Tamil Nadu, journalist Priyanka Shankar described both the benefits that wind farms had brought to two districts of Tamil Nadu, Tirunelveli, and Kanniyakumari and some of the problems that their haphazard development created:

> Muppandal, a booming town situated in the heart of this region, is home to India's largest operational onshore wind farm in terms of installed capacity. The farm features turbines from many private players, and was developed by the Tamil Nadu Energy Development Agency (TEDA). The wind turbines here are some of the oldest in the country, and range from 200 kilowatts (KW) to 1,650 KW. What makes Muppandal a great site to install wind turbines is its location, which receives uniquely powerful winds from the Arabian Sea. The wind turbines have transformed the lives of the people of the region: 'Wind is a huge boon for the youth of south Tamil Nadu,' declares Thangaraj, standing next to an 88-metre wind turbine that makes a heavy sound as the blades tear through the air . . . 'My father is not educated, and my mother passed away due to cancer; but they gave me education. I completed engineering and I am now working in a job that I am proud of. I live in a pollution-free atmosphere, and I am doing a green job.' Sooriya Priya, 27, whose husband worked on a wind farm echoed his sentiment: 'There can be positives and negatives of large-scale renewable energy projects, but we get continuous power now. We have no power cuts except the maintenance cuts. This is a huge relief for us. We feel secure and our lifestyles have improved.'[5]

Wind farms have also created large numbers of permanent jobs for the youth of these areas. For those who have received a secondary or technical education the wind turbines have come as a blessing because they have created large numbers of the kind of jobs for which they have been trained and that are available close to their homes. This has eliminated the overhead costs they have been incurring travelling to cities to find the jobs for which they have been qualified.[6]

But as the crowding of wind turbine locations as indicated in Figures 14.1 and 14.2 shows, the absence of a masterplan for setting up wind turbines in the region as a whole has set off an uncontrolled proliferation of windmills. This has resulted in suboptimal power generation in several places. To prevent this, turbine owners have begun to fence in the wind coverage area around their wind turbines, buying or leasing some of the land from its owners where necessary. These fences are eroding the popularity of the wind turbines among the villagers, as they are not only beginning to impede the free movement of people through the fields, but also preventing them from grazing their cattle on hitherto unfenced lands. The lack of a regulatory policy for setting up wind farms is therefore beginning to create a backlash that can reduce the area available for wind power generation and slow down the shift out of fossil fuels in the future.

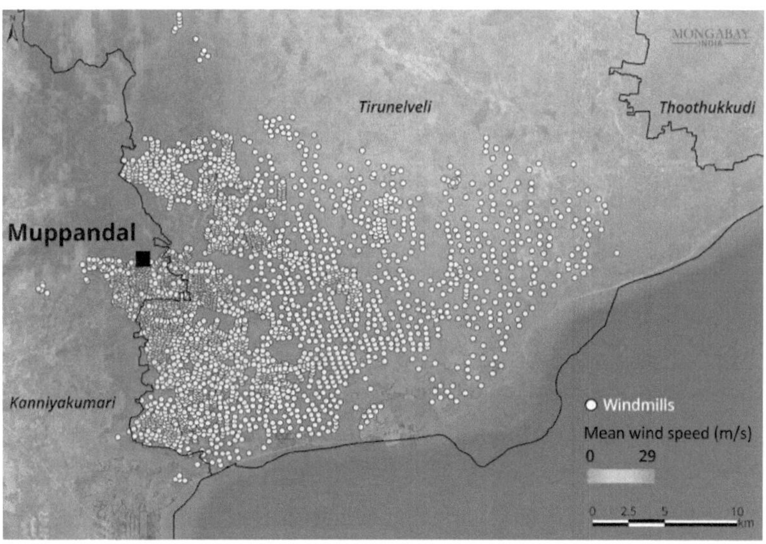

Figure 14.1 Wind turbines in Tirunelveli and Muppandal districts of Tamil Nadu.

Source: TechforWildlife.com. https://india.mongabay.com/2022/04/the-wind-farm-paradox-in-southern-tamil-nadu/

Figure 14.2 Crowding of wind turbines.
Source: *Business Standard*. Wind energy has huge potential in India, to cost less by 2026: GWEC. https://www.business-standard.com/article/economy-policy/wind-energy-has-huge-potential-in-india-to-cost-less-by-2026-gwec-122083001400_1.html (viewed 23/9/2022).

Solar and wind power generation have been seen as ends in themselves, not as means to an end.

At present, despite the rapid spread of wind farming in a few states, there is virtually no awareness in the Government of India of how far harnessing solar and wind power can go towards eliminating poverty, unemployment, and farmers' suicides in rural India. The sole aim of the government's policies so far has been to reduce its cost of generation in relation to coal- and gas-based power. The United Progressive Alliance government that ruled India from 2004 to 2014 had adopted the policy in vogue in most countries wishing to encourage renewable energy development of giving the producers a feed-in tariff significantly higher than the price of competing fossil-based power and locking this in for a negotiated period of time, usually 15–20 years. When he was the chief minister of Gujarat, Prime Minister Modi used this policy to increase Gujarat's solar and wind energy generation capacities, offering as much as Rs 5–15 per kwh to attract investment. As a result, in 2014, Gujarat accounted for 40% of India's solar generation capacity and in 2015 the state became the largest producer of wind energy.

However, after becoming Prime Minister in 2014, Modi raised the renewable energy target from 100 GW to 175 GW by 2022, and, possibly in response to complaints by the US Trade Representative against the minimum domestic content requirement for renewable energy investment in India, the government changed its policy and began to award contracts based solely on reverse auction tenders for wind and solar power projects. Reverse auctions involve obtaining bids from many firms, but since the government is the only buyer, it is able to drive down bid prices to the lowest sustainable level. This policy shift led to a crash in the price of both wind and solar power. Furious competitive bidding by domestic and international firms brought the former down to Rs 2.43 per kWh in 2017 and to Rs 1.99 per kWh for the latter in December 2020.

These tariffs proved unsustainable. In 2019, only 2.3 GW of wind capacity was installed—nearly half of the 4.1 GW installed in 2017. Between 2017 and 2020, 17 GW of capacity was auctioned across the country by various power purchasing agencies, but nearly one-third went unsubscribed or had to be renegotiated post-award due to stringent tender conditions. The winners of existing contracts began to withdraw from their projects, preferring to cut their losses. Others renegotiated their contracts. As a result, 80% of the projects were delayed by 6–12 months.

The agreed tariffs eventually settled around an average of Rs 2.80 per kWh for wind and Rs 2.40 per kWh for solar power.[7] But this created fresh problems when the state governments began to use these as tariff caps for the bids that they would entertain. The result was that the next three auctions failed to attract any bids. As a result, the addition of wind and solar generating capacity slowed to a crawl. An almost unnoticed consequence of the reverse auction tendering process was that it also ended whatever chance India had of setting up a domestic manufacturing base in renewable power generation. In 2018, the Modi government tried to compensate for this by raising tariffs on imported components, but since the reverse auction process remained in place, this increased the disadvantage of domestic component manufacturers against turnkey bidders by increasing instead of reducing their returns, increasing thus their cost of production disadvantage.[8]

The price compression caused by the reverse auction process also forced the winners of contracts to exploit every available method to reduce their costs. As the Observer Research Foundation (ORF) study, chapter 8, showed with respect to the plight of the owners of land acquired for solar power projects, the main victims have been the already dirt-poor owners of land and the cattle-herders who previously had free access to it for grazing. Among these groups, opposition to solar farms, in particular, has begun to grow.

In January 2023, the government conceded the failure of its earlier policy and scrapped reverse auction bidding for wind energy.

Finding the land for solar and wind projects

Other than the rock-bottom electricity tariffs they have been forced to accept, investors have faced three other obstacles: the difficulty of acquiring land; the reluctance or inability of state power grids to accommodate renewable energy by backing up coal- and gas-based power stations during the hours when wind and solar power is available; and the tariff caps mentioned above that leave them to bear the entire burden of future increases in the cost of generation. Of the three, the availability of land is almost certain to prove the biggest hurdle. Up to now, whenever the central or a state government has needed land for a public purpose, it has simply acquired it, i.e. taken it away from its owners or, in the case of forest and tribal lands, from those who have enjoyed customary usership rights, by a process of 'land acquisition for a public purpose'. The law governing this was first enacted by the British in 1894, amended many times, and eventually replaced by the *Land Acquisition, Rehabilitation and Resettlement Act, 2013*. The 2013 Act sharply increased the compensation given to the owners and customary rights holders, but did not change the expropriatory nature of land acquisition. This was because the compensation offered remained based, as in the original Act and all of its subsequent amendments, upon the value of the land in its *present* and not its *intended future* use. This clause therefore continued to systematically exclude the poor from the benefits of the economic development that the change of land use would bring. What is worse, since most of the compensation was given in monetary terms, its value was steadily eroded by inflation. Over seven decades, the further impoverishment of the already poor has been so devastating that few people, anywhere in India, now give up their land for development, no matter how important the stated public purpose, without first fighting a lengthy political and legal battle.[9]

The contestation over land rights is not new in India. As a follow-up study by ORF of land use in solar and wind power projects has pointed out, the expansion of renewable energy power generation capacity targets is causing concern over the availability of land and related issues for developing renewable energy projects. For project developers, the key concerns are the availability of suitable land and the difficulty of acquiring vast tracts of land from a number of poor rural title holders. For the policy makers the challenge has been to find the right balance between meeting targets for decarbonization, ensuring food security, and finding viable alternative livelihoods for those displaced from the land for renewable energy development. For poor landowners, problems include obtaining fair compensation for land appropriated for renewable energy projects and relocating to areas where they can find alternative livelihoods. Official estimates suggest that over 60 million people were displaced by development projects and less than a third were resettled.[10]

The opening line of the chapter on land (chapter 3) of India's 12th Five Year Plan (2012–17) observes that 'India has a long history of social discrimination closely linked with denial of access to land'. Land acquisition from the rural poor for large industrial and mining projects is in general skewed in favour of industrial developers.[11] What is difficult to understand is why, in more than seven decades of independence, no central or state government has ever considered making the poor—the scheduled castes and tribes, and the growing army of small and marginal farmers created by the subdivision of holdings—partners in their economic development plans instead of its victims.

The potential of SPV power to generate incomes and jobs and its impediments

In sharp contrast to wind turbines, which require little actual land and therefore relatively little change of land use, SPV is the most land hungry of the renewable energy alternatives faced by India and the world because it requires the whole of the land from which it draws its power to be taken out of alternative uses. But it compensates for this by generating a larger amount of electricity. Various projects in India have shown that ground-mounted SPV power plants generate 25–30 MW/km^2 of land, depending on topography, latitude, insolation, and ambient temperature.

Assuming conservatively that it is only 25 MW and that the average insolation is 1,600 hours, a regular royalty payment of Rs 0.30 per kWh of power generated from their land would still provide an annual revenue for the land owners amounting to 25,000 (kWh) × 1,600 (h) × 0.30 (royalty per kWh) = Rs 12 million. When distributed among the land owners, this would provide them individually with an annual income of Rs 120,000 ($8,000) per hectare of land covered. Not only would this change their lives forever, but also, since it would increase the purchasing power of those with the highest propensity to consume, it could trigger up to a five-fold or greater increase in the GDP of the area.[12]

Figure 14.3 Solar panels.

Source: David Monniaux. https://commons.wikimedia.org/wiki/File:Mafate_Marla_solar_panel_dsc00633.jpg.

As was pointed out in Chapter 13, India has 254,000 km^2 of wasteland in the eight semi-arid states of the country, of which the single largest chunk, measuring approximately 160,000 km^2, is in the Indian portion of the Thar Desert in Rajasthan. By 2050, if half of this wasteland were brought under development for utility-scale SPV power plants, it could generate Rs 1.428 trillion ($17.85 billion) of royalties for the owners and customary rights holders of this land, who are among the poorest of the poor in the country. Assuming a multiplier of 5, the overall increase in GDP generated by these royalties on the land developed for SPV would likely be significantly more than four times the initial expenditure, greater than the average multiplier for the Indian economy as a whole.

Unfortunately, the pressure that the reverse auction bidding process has put on investors in recent years has ensured that they will leave no stone unturned to minimize the transport and the infrastructure development portion of their costs. To do this, they have looked to state governments to acquire the land they need for their solar farms. As the first of the two ORF reports, which was cited earlier showed, since land and infrastructure development costs are lower on inhabited land, only 11–12% of the land brought under SPV plant development has been virgin. The balance has been acquired from existing owners by the state governments at prices decided by them under the rules laid out in the 2013 Act, backed by promises of social, educational, and infrastructural development that have seldom been honoured.

Royalties on the coverage area of wind turbines can have an equally important impact on rural poverty

Because it requires so much less land, the harnessing of wind power may not have as dramatic an impact on the income of the poor as the tapping of solar power can have under the conditions described above. But if India's state governments were to adopt a modified version of the 'Good Neighbour' payments policy described in Chapter 13, it could also transform the lives of the desperately insecure poor in the semi-arid areas of the eight dryland farming states which have the bulk of India's wind power generating potential and reduce the spate of suicides in them to a trickle. The governments of the eight wind-rich states could get an idea of the transformation this could bring by studying what took place in the Great Plains states of the US. This was described by Elizabeth Wiese in an article published in 2020 in *USA Today* as follows[13]:

> In an increasingly precarious time for farmers and ranchers, some who live in the nation's wind belt have a new commodity to sell—access to their wind. Wind turbine leases, generally 30 to 40 years long, provide the landowners with yearly income that, although small, helps make up for economic dips brought by drought, floods, tariffs and the ever-fluctuating price of the crops and livestock they produce. Each of the landowners whose fields either host turbines or who are near enough to receive a 'good neighbor' payment, can earn $3,000 to $7,000 yearly for the small area—about the size of a two-car garage—each turbine takes up.

The median income in Cloud County is about $44,000, according to the 2018 US Census. For Tom Cunningham who's been farming between Glasco and Concordia, Kansas for 40 years, the Meridian Way Wind Farm income has made an enormous difference. Cunningham's lease payments allowed him to pay off his farm equipment and other loans.

Before the wind turbines, things were rough, he recalled. Depending on the national and international economy, some years he broke even, some years he made money, and for more years than he cares to think about he was on the edge. He had to take a job in town to make ends meet and for a time was what he calls 'functionally' bankrupt. 'This isn't money that other people would think is very much,' he said. 'But it made an enormous difference to us.'

In central and southern India the impact on the lives of the poor of such a royalty payment on electricity generated from wind turbine installations may not be as dramatic as it would be in the arid and wastelands most suited for solar power generation. But it too will transform their lives. This is because in addition to the cash income over and above what they already earn from farming, wind power utilities are likely to create a much larger number of jobs than would utility-scale SPV power plants.

In 2020, electricity tariffs in Tamil Nadu ranged from Rs 4.50 per kWh for up to 300 kWh per month to Rs 11 per kWh for consumption above 1,000 kWh. The price of electricity supplied by the state electricity boards across the country was more or less similar and averaged Rs 6.15 ($0.082) per kWh. Table 14.1 shows India's year on year installed wind power, annual wind power generation, and annual growth in wind power generation since 2006. At a CUF of 21%, which has been the average for the past 15 years, a 3-MW wind turbine, such as the Vestas V90, could generate 5,518,800 Kwh of power in a year. A 5% royalty (Rs 0.31 per kWh) added to this tariff would be barely noticed by the paying consumers, but each turbine would generate Rs 1,710,000 in royalties every year for distribution among farmers within its coverage area. Divided among the owners of the 44.4 hectares of land (the average coverage area estimated by the NREL for the Vestas V90 turbine in the US), this would add Rs 38,530 per hectare ($480 per hectare) to the net cash earnings of each farmer.

In the rain-dependent farmlands of both Tamil Nadu and Karnataka, the small and marginal farmers each own an average of 1.4 hectares of land. Thus, falling within the coverage area of a 3-MW wind turbine would add Rs 53,900 to each family's annual cash income. This assured annual revenue would begin a rural transformation that can go well beyond meeting the immediate needs of the rural poor, for it will provide a cushion to small and marginal farmers that will shield them from unforeseen setbacks such as a drought, accidents, or a sudden illness in the family. And, as in the US, it will provide an assured commitment, a guarantee against which they will be able to obtain loans from the banks to improve the lives and education of their children.

In March 2021, the installed capacity of wind turbines in India was 38.785 GW and was growing at close to 2 GW a year. Although the most recently installed wind turbines had a generating capacity of 3.6 MW, the vast majority still had nameplate capacities of 1.5–2 MW and had been installed at hub heights of 80–100m. As Table 14.1 shows, the CUF of the wind power industry had stabilized during the previous 4 years at a little less than 20%.[14] [15] When the potential of the eight rain-dependent farming states is fully exploited, then—assuming conservatively (1) that the current average CUF has risen by 25% because of greater hub heights (Table 14.1); (2) that by 2050, half of the 120 mha of rainfed farmland and all of the land devoted currently to permanent fallow identified in Chapter 13 as suitable for wind farming has been brought under development; (3) that by 2050, the average generation capacity per turbine has risen to 3 MW; and (4) that two such turbines are installed on every square kilometre of the above land—we may conclude that these states

Table 14.1 Installed wind power capacity, generation, and capacity utilization factor since 2006.

Financial year	Installed capacity (MW)	Generation (GWh)	CUF(%)
2006–07	7,850		
2007–08	9,587		
2008–09	10,925		
2009–10	13,064		
2010–11	16,084		
2011–12	18,421		
2012–13	20,150		
2013–14	22,465		
2014–15	23,447	28,214	13.7
2015–16	26,777	28,604	12.2
2016–17	32,280	46,011	16.3
2017–18	34,046	52,666	17.7
2018–19	35,626	62,036	19.9
2019–20	37,669	64,485	19.5
2020–21	38,785	59,824	17.6
2021–22	40,355	68,640	19.4

Sources: India Renewables Dashboard. https://www.renewablesindia.in/ (viewed 27/9/2022); Central Electricity Authority: Monthly Renewable Energy Report March 2022. https://cea.nic.in/wp-content/uploads/resd/2022/04/Broad_overview_March_2022.pdf (viewed 29/9/2022).

could generate 7,884 TWh of power, and a 5% royalty on the current price of electricity would add Rs 2.44 trillion ($30 billion) to the income of the poorest, most insecure people in the country.

Notes and References

1. Climate Change 2022: Mitigation of Climate Change. Working Group III. Contribution to the Sixth Assessment Report of the Intergovernmental Panel on Climate Change: Summary for Policymakers. https://www.ipcc.ch/report/ar6/wg3/.
2. UNFCCC, COP 27, Sharm el-Sheikh Implementation Plan, 2022. https://unfccc.int/documents/624444 (viewed 21/9/2022).
3. So far, none of the countries that have set themselves target dates for ending the use of fossil fuels by 2050 have taken even the first steps towards doing so. An IMF

estimate has shown that the world is spending $5.9 trillion dollars on repairing the damage that global warming is doing to the world through an increasing frequency and severity of extreme weather events. IMF: Fossil Fuel Subsidies, 2020. https://www.imf.org/en/Topics/climate-change/energy-subsidies (viewed 21/9/2022).

4. UNFCCC, COP 27, Sharm el-Sheikh Implementation Plan, 2022. https://unfccc.int/documents/624444 (viewed 21/9/2022).
5. Priyanka Shankar: Lands sold, jobs gained, and a new way of life. India Together, 26 April 2022. https://indiatogether.org/wind-tn-environment (viewed 22/9/2022).
6. Ibid.
7. Arjun Joshi: Five Lowest Solar Tariffs in 2022, 30 December 2022. https://mercomindia.com/five-lowest-solar-tariffs-in-2022-infographics/ (viewed 23/9/2022). 2022 saw a steady rise in auction bid prices from Rs 2.29 per kWh to Rs 2.66 per kWh.
8. Behuria, Pritish (2020). The politics of late development in renewable energy sectors: dependency and contradictory tensions in India's National Solar Mission. *World Development* 126, February 2020, 104726. https://www.sciencedirect.com/science/article/pii/S0305750X19303754 (viewed 26/9/2022).
9. The opening line of chapter 3 of India's 12th Five Year Plan observes that 'India has a long history of social discrimination closely linked with denial of access to land'. Land acquisition from the rural poor has in general been skewed in favour of industrial developers. Official estimates suggest that over 60 million people were displaced by development projects and less than a third were resettled. Shankar, Priyanka (2022). Lands sold, jobs gained, and a new way of life. *India Together* 26 April 2022. https://indiatogether.org/wind-tn-environment (viewed 26/9/2022).
10. Lydia Powell, Akhilesh Sati, and Vinod Kumar Tomar: Solar Energy in India: Contestation of Land Rights, 3 May 2022. https://www.orfonline.org/expert-speak/solar-energy-in-india/ (viewed 26/9/2022).
11. Reddy, I.U.B. (1993). *Displacement and Rehabilitation*. Mittal Publications. https://mittalbooks.com/products/displacement-and-rehabilitation.
12. In India, the 'multiplier' (a concept developed by the British economist J.M. Keynes to describe the number of times any given expenditure changes hands, adding to the income of subsequent recipients, before all of it is absorbed into savings) is currently a shade under 4. For the very poor, whose propensity to save is very low, it will be considerably higher.
13. Elizabeth Weise: Wind energy gives farmers a new crop to sell in tough times. https://eu.usatoday.com/story/news/nation/2020/02/16/wind-energy-can-help-american-farmers-earn-money-avoid-bankruptcy/4695670002/ (viewed 27/9/2022).
14. Central Electricity Authority: Monthly Renewable Energy Report March 2022. https://cea.nic.in/wp-content/uploads/resd/2022/04/Broad_overview_March_2022.pdf(viewed 29/9/2022).

15. Table 14.1 shows that in 2018–19, the CUF has averaged a steady 19.9%. However, since the average increase in generation capacity was around 2 GW a year, it should be assumed that the most recently added turbines have generated electricity for only a part of the year. That assumption pushes up the CUF of operating plants to 21%. If by 2050, all but a few of these have a hub height of 120 m, that could push the CUF up to close to 25%.

15

Combining Decarbonization and Economic Growth to Open a New Future for the Rural Poor: Biomass gasification could be the answer

As we pointed out in Chapter 5, agriculture in the Indo-Gangetic Plain is facing the very opposite of the crisis confronted in the drylands of the south, centre, and west. While farmers in the drylands are beset by chronic scarcity, their peers in the Indo-Gangetic Plain are suffering from chronic overproduction. While the farmers of the drylands are in acute need of protection from the inclemencies of weather, the farmers in the irrigated areas are in a desperate search for ways to break out of what was initially a 'cereals trap' and has now become a 'cereals and vegetables' trap. The way this has affected their livelihoods has been described in detail in Chapter 5.

Horticulture has been the final refuge of the ever-increasing number of small and marginal farmers created by the subdivision of holdings passed to each succeeding generation. But in the absence of cold storage in their villages or in nearby locations, this too has become a trap in which the more they produce, the poorer they become. We described how small and marginal farmers who have turned to horticulture to improve their livelihoods have become victims of the market rather than its beneficiaries, reflecting the perishability of their products and the absence of cold storage in the villages where they could store their crops. We offered several examples of the vast gap between the farmgate prices the farmers receive and the final table value of their produce in the cities. An estimate of how much the perishability of their produce costs them may be inferred from the fact that although horticulturists produced 12% more

Sun, Wind, and Biomass. Prem Shankar Jha and Michael B. McElroy, Oxford University Press.
 DOI: 10.1093/9780198944768.003.0015

fruit, vegetables, and flowers (326 Mt) than cereal farmers produced foodgrains (291 Mt), they earned only half as much.[1] By contrast, the year-round average table value of vegetables in the cities corresponds to three times the price of cereals. Most of this six-fold difference in earnings can be traced to the absence of cold storage in the villages. Just how much difference a single small cold storage facility in a village could make is revealed by the following description of how two, very small cold storage units have transformed the future of two remote villages, one in Karnataka and the other in Gujarat.

In April 2023, for the first time, the Kunbi tribal community in Joida (Karnataka) harvested their crop of pickling mangoes without a trace of nervousness. This was because they now had a cold storage unit right in their village where they can keep their mangoes before taking the fruit on a long journey to the nearest market. These mangoes, a local wild variety called *appe midi*, have to be harvested at a tender stage, otherwise they lose their value. Joida, a predominantly Kunbi tribal *taluk*, is very remote and doesn't have basic facilities. In May 2022, 27 villages got electricity (from the Karnataka State Electricity Board) for the first time, but only for 1 day. This same *taluk*, with so little to speak of, now has a smart cold storage. It was gifted to the villagers by CoolCrop Technologies (a small start-up company formed in 2017 by an Indian technologist, Neeraj Marathe, with four colleagues from India and the US). This company set out to provide storage solutions for small growers in remote areas and established its first solar-powered Cold Storage unit in Kunbi. Built at a cost of $5,000, this was a low-cost machine (with solar panels) mounted on a roof of asbestos sheets and attached to the house of a young farmer, Ghanshyam Derekar, and his brothers.

Joida is famous for its production of a big tuber called *mudli*. During the tuber harvesting season in November and December, the *mudli*, which is much prized by the Saraswath Brahmin community, sells for Rs 70 per piece. By February, the price of the same tuber rises to Rs 120. Joida can now store 500–600 *mudlis* and earn an additional income of Rs 25,000 while waiting for the price to escalate during the off-season. That only will be the beginning. According to the report, apart from storing their *mudlis* and their prized *appe midi* mangoes, the farmers of the village now plan to grow and store tomatoes and other vegetables and keep the cold storage unit full for 10 months of the year.[2]

Joida's mini cold store, with a capacity of approximately 4 tonnes, was CoolCrop's first venture. Built in 2018, it began to operate in January 2019. Joida is one of the poorest and remotest villages in Karnataka. Yet this one small cold store has been able to increase the villagers' incomes by a quarter or more. By December 2020, CoolCrop had contracted to build similar cold storages in 13 more villages. Its second plant, in Kakadkheda, another remote village in Gujarat state, went on line in 2019. According to a report carried by the German TV channel *Deutsche Welle on this*, this cold store has raised the income of the farmers in Kakadkheda by 30%. According to a report carried by the German Television channel, Deutsche Welle in 2022, Marathe and his team intended to set up 200 such units over the succeeding 5 years.[3] In 2022, therefore, CoolCrop was well on its way to proving that a network of village-based cold storage units could transform the future of the poorest farm families in the country. But relying on SPV power alone will not solve the problem of perishability, for it will have to be backed up by battery storage for several hours a day even in the more developed states of the country. This could make it prohibitively expensive for all but the most highly valued produce.

Two-stage biorefineries

There is, however, another way to ensure a continuous supply of power to cold storage units that could transform the future of rural India. This is to gasify crop residues, mainly paddy and wheat straw, and maize stover, in air-blown gasifiers in the village itself, and use the resulting lean fuel gas to ensure a continuous supply of power for cold storage units, to produce briquettes, and to sell the remaining biochar to steel plants as a substitute for coking coal or to biorefineries for conversion into transport fuels or green hydrogen as the market dictates.

The impact that biofuels can have, was discussed briefly in chapters 13 and 14. In brief, the transport sector currently accounts for 13–14% of CO_2 emissions, but this is likely to rise, in a business-as-usual emissions scenario, to 30% by 2040. Biofuels, electric vehicles, and hybrids can reduce this source drastically over the next two decades and completely eliminate it well before 2070. *But only the production of biofuels through crop residue gasification can do this and simultaneously rejuvenate the rural economy.* It can do so because, while it will add immense value to the straw that the farmers now consider a waste product, it will also simultaneously provide the electricity for the village cold storage units that fruit and vegetable growers have so far been denied. India's pressing need to replace imported coking coal and crude oil not only to decarbonize energy consumption but also to reduce India's presently crippling dependence on imports of both will ensure that farmers get the best price for their biochar briquettes that the market can afford.

Social benefits from two-stage biorefineries

The following study of the difference that two-stage biomass gasification can make to the village economy in the state of Punjab shows how it can transform the future of rural India. In 2020, Punjab had a rural population of 27.7 million living in 12,278 villages, cultivating 4.2 mha of land. Of this, 3.52 mha grow wheat alone and 3.10 mha also grow paddy. The area under horticulture is 381,000 hectares, of which 273,000 hectares grow vegetables and 86,000 hectares produce fruit.[4] The remaining 600,000 hectares are devoted to sugar cane, cotton, floriculture, and various spices. The average cultivated area in each village is 342 hectares. Of this, 252 hectares grow both wheat and rice, 34 hectares grow wheat and one other crop, and 23 hectares grow vegetables. The remaining 33 hectares grow fruit, flowers, sugar cane, and cotton. The average yield of paddy was 4.36 t/ha in 2019, the average yield of wheat was 5 t/ha, and that of vegetables just under 20 t/ha. This Table 15.1 is for 2015–16.

Punjab's 4.2 mha of cultivated land are farmed by 938,000 families. Of these, 207,000 are marginal farmers who cultivate an average of 1.4 hectares each. Most of the vegetables are grown by them. A further 368,000 families are classified as 'small farmers' who each cultivate 2.67 hectares and 305,000 are 'medium farmers' with holdings averaging 5.67 hectares. The 673,000 farmers in these two categories, who till 70% of the land, are the backbone of Punjab's agriculture. Their average holding is 4.12 hectares and there are an average of 55 such families in each village. Since this is also more or less the average size of operational holdings in the whole of Punjab, it is reasonable to assume that most of their land is devoted to

Table 15.1 Size distribution of farmland in Punjab.

Size	Number of farms (thousands)	Area (thousand ha)	Average per farm (ha)
Marginal	207	290	1.40
Small	368	984	2.67
Medium	305	1730	5.67
Large	58	857	14.77
Total	938	3861	6.12

Source: India: Census of Agriculture 2015–16.

double cropping of rice and wheat. Another 20 families depend mainly, if not entirely, on vegetable farming. Finally, there are 4 to 5 rich farmers in every village cultivating 60–75 hectares of land. These individuals have the best contacts and the greatest access to the banking system. So they are the ones who have branched out into fruit orchards, sugar cane, and, more recently, aggressively into cotton. The area sown with cotton in a single year increased from 249,000 hectares in 2018 to 388,000 hectares in 2019. This is about 40% of the land owned by the rich farmers.

Identifying the principal beneficiaries of biomass gasification requires us to examine its impact at the village level. Wheat and rice generate approximately 6 and 5 tonnes of straw per hectare respectively.[5] The average village therefore generates 11 tonnes of straw per hectare from its 252 hectares of double cropped land and probably 8 tonnes, or thereabouts, of crop residues from the remaining 90 hectares devoted to cultivation of wheat and miscellaneous other crops. The straw, maize stover, cotton stalks, and other crop residues available in a village add up therefore to more than 3,400 tonnes.[6] Gasification of this straw in simple air-blown gasifiers would yield close to 1,000 tonnes of biochar and more than enough lean gas[7] to supply cold storage units with the power they are unable to get from the state electricity grids. After removing ash and some other soluble impurities from the biochar, there will be close to 800–900 tonnes of clean biochar containing 80–90% of carbon and no sulphur. These briquettes can be sold to steel plants to replace imported coking coal in their blast furnaces or to biofuel plants as a high-grade feedstock for producing any transport fuel the economy needs.

Coking coal for steel plants

India has very little coking coal in its otherwise ample reserves of coal and is importing therefore 80% of its requirement, mostly from Australia. Imports have grown from 44 Mt in 2015–16 to 75 Mt in 2020–21 and are expected to rise to 160 Mt by 2030.[8] Since the world is also running out of coking coal, its free on board (FOB) price has risen from $80 per tonne in 2015–16 to $200 per tonne in 2019–20. Reflecting this increase, the market price of coking coal in the Indian market in November 2018 was Rs 23,000 ($305) per tonne. A year later it had risen to Rs 27,000 ($360) per tonne. After the start of the war in Ukraine f.o.b. (Freight on Board) coking coal prices rose sharply to $341 per tonne and then seesawed between that and a relative low of $212 in the third quarter of 2024.[9] In November 2023, the market price of coking coal in India was $350 per tonne.[10]

These market prices virtually guarantee a 'village-gate' price of Rs 12,000–15,000 ($160– 200) to the farmers. Thus the 55 'median' land-holding families in the average Punjab village with just over 4 hectares of land that presently earn Rs 264,000 in a normal year will gross an *additional* Rs 150,000 or thereabouts from the sale of approximately 12 tonnes of cleaned biochar briquettes obtained from their 45 tonnes of paddy and wheat straw. When a cold storage facility is developed in their village or in a nearby village they may earn an additional income from the sale of lean gas.

Increase in family income for median farmers

In 2015–16, the average 4-hectare-owning farm family earned a net income after meeting the cost of hired labour and all bought inputs of Rs 45,000 or thereabouts per hectare from wheat and Rs 21,000 per hectare from paddy.[11] The income from double cropping its land, net of paid out costs, was therefore Rs 264,000. The sale of biochar briquettes produced from rice and wheat straw gasification could thus increase their net annual income by 45–56%. For the 673,000 families with a median land-holding, this would mean an increase of Rs 80–101 billion ($1.15–1.4 billion).

For marginal farmers cold storages could be the salvation

In 2018, Punjab as a whole had a total of 378 cold storage units. By mid-2020, this number had risen to 402. Of the 378 in 2018, 362 stored only potatoes and 7 stored dairy products. The balance accommodated a variety of foodstuffs, but all of the 378 units were in or at the edges of the state's main towns and cities. There were, and still are, no cold storage facilities in the rural areas due to limitations imposed by 8-hour power cuts and fluctuating voltage levels, which are the norm in Punjab, as they are throughout the country.[12] As a result, villagers use electricity only for running their tube wells, for lighting, and for powering their fans, even when power is available. Not surprisingly, therefore, the average Indian village has a total electricity demand of a paltry 1,826 kWh/day, with about 52% consumed by households and only 7% by enterprises. The remaining two-fifths are used by agriculture, almost entirely through the application of diesel generators to run irrigation pumps. This is the lacuna that power generation from lean gas could fill—for every 1,000 m^3 of lean gas contains the energy equivalent of 28 gallons of recoverable liquid fuel.[13]

Impact on rural incomes

Punjab produces a seventh of the total rice and wheat grown in the country and therefore a seventh of the rice and wheat straw. So it is possible to extrapolate the income that could be generated by gasifying these residues proportionately. At a rough estimate, rice and wheat straw gasification alone could increase net farm incomes by Rs 700–800 billion. Add cotton stalks and maize stover and the source could well exceed Rs 1 trillion ($13 billion). This, however, is only a quarter of the increase in GDP that can be anticipated, because the injection of this money into the economy will create more income. The ratio of the initial injection to the total income that it would generate is the inverse of the savings rate in the economy. In India, the average savings rate is 25%, so the 'multiplier' is close to 4. The resulting increase in GDP could thus be in the order of Rs 4 trillion, i.e. 2% of India's GDP in 2019–20.

Impact on rural employment—a simplified model

This would represent a significant addition to GDP, but what is far more important is that the increase will take hold where the social impact will be the greatest. It can be anticipated to create permanent non-agricultural jobs that will take the pressure off the land and increase the income security of the least prosperous families in the rice and wheat growing states. The operation of a gasifier, a briquetting plant, and a cold storage unit with the capacity to store up to 400 tonnes of fruit and vegetables would create three-shift employment in the gasifier industry and two-shift employment in the cold storage and briquetting plants. Between supervisors, technicians, transport workers, and manual labourers and allowing for days off, leaves, and sickness, it is difficult to see how the full utilization of crop residues will not create 50 or more jobs in each village. That would mean 600,000 or more full-time jobs in the state as a whole.

The impact this will have can be gauged from the existing pattern of employment in Punjab. In 2017–18, 62% of Punjab's approximately 30 million population,[14] i.e. 18.6 million persons, lived in its rural areas in 3.325 million 'households', which contained 4.32 million nuclear families. Of these nuclear families, 1.95 million were self-employed: two-thirds functioned as farmers and a third as traders, transporters, and small entrepreneurs.

Another 860,000 families, 19.9%, had one or more wage- or salary-earning members. But more than a million families depended on 'casual', i.e. seasonal or sporadic, work. By the definition of employment used in the industrialized countries, the breadwinners in these families are unemployed, for they have no regular jobs, no fixity of tenure, no social security, and no old age insurance. The 600,000 permanent jobs that gasifiers and cold storage units could generate would be secure and more or less permanent. They could therefore cut the number of families living in acute insecurity by as much as two-thirds.

But that will only be the initial impact. As the 'multiplier' effect of the initial increase in farm incomes works its way through the economy, it will create four times as much additional expenditure and therefore up to a similar number of additional jobs.[15] In all, therefore, a gasifier, a cold storage unit, and a briquetting plant in each village in Punjab could create close to 2.5 million additional, regular, salaried jobs. Most of these will be within the state. And two-thirds or more, i.e. 1.6 million or thereabouts, are likely to be permanent, with participants earning a regular income.[16]

The menace of drugs

In Punjab, apart from increasing the security of life and raising the standard of living, this increase in salaried, non-agricultural jobs will go a long way towards mitigating the most serious social problem that the people of the state are facing today. This is the absence of non-farm jobs for the sons of the Green Revolution generation of farmers, whose children have gone to schools and colleges and whose sons, in particular, no longer want to work in the fields. The absence of alternative employment has driven large numbers into drug addiction and crime. The ready availability of secure, relatively well-paid non-farm jobs in the villages would go a long way towards lifting this scourge.

Extrapolating from Punjab to India

Punjab grows one-eighth of India's foodgrains and is responsible therefore also for one-eighth of the crop residues derived from foodgrains. On the assumption that the ratio between crop residues gasified, income, and employment generated remains the same in the other parts of the country, biomass gasification could generate 20 million or more permanent non-agricultural jobs in the rural areas. The multiplier effect of this investment could create two to three times as many more jobs locally in construction, trade, and associated manufacturing. Most of these jobs will be sited in the Indo-Gangetic Plain and the rice-growing riverine and delta regions of southern and western India. These are precisely the areas that wind and solar power cannot reach. When compared with India's estimated population of 1.4 billion in 2017, this does not seem like a large number, but its true impact will be the effect it would have on the lives of the 44 million rural households in the country that are surviving currently on casual labour, supplemented by remittances sent by relatives who have migrated to the cities in search of work. This number will initially be cut in half and then reduced further as the multiplier effects kick in. As in the case of wind farms in the drylands, for these families, an even more important benefit will be the security of knowing that there are non-farm jobs close to home that their children will be able to take up when they finish their schooling and higher studies. As pointed out in Chapter 3, citing the studies by University of Bath professor Santosh Mehrotra, these are the jobs that their children feel entitled to and for which they hanker. Finding them close to home will change their lives.

Notes and References

1. Estimate culled from constant price data given in Government of India, Ministry of Agriculture & Farmers' Welfare: Horticultural Statistics at a Glance 2018.
2. Shree Padre: Cold storages for villages: a mini answer from CoolCrop. *Civil Society Magazine*, 28 February 2019. https://www.civilsocietyonline.com/business/cold-storages-for-villages-a-mini-answer-from-coolcrop/ (viewed 6/12/2020).
3. Cool Storage in India. https://www.dw.com/en/cool-storage-in-rural-india/av-52186180 (viewed 4/10/2022).
4. The rest is under maize, sugar cane, cotton, floriculture, and spices.
5. Studies cited here estimate that the paddy to straw yield varies from 1.13 to 1.3 t/ha depending on the type of paddy grown. Singh, R.B., et al. (1995). Rice Straw—Its Production and Utilization in India. https://www.semanticscholar.org/paper/Rice-straw%2C-its-production-and-utilization-in-Singh-Saha/839e9194d8ee2c73967e43acc913d79d80cd9948. Singh, Kiran and Schiere, J.B. (eds) (1995). *Handbook for Straw Feeding Systems.* ICAR. Kaur, Arjinder (2017). Crop Residue in Punjab Agriculture—Status and Constraints. Department of Economics and Sociology, Punjab Agricultural University. *Journal of Krishi Vigyan* 5(2):22–6. http://iskv.in/wp-content/themes/iskv/volume-pdfs/8296866dd23487ae0a38ee40a27f2a26pages_28-32.pdf (viewed 4/10/2022).
6. Sugar cane leaves, tops, waste, and bagasse are not included here because these are available only at the sugar mills.
7. With a calorific value of 1,500–1,600 kcal/Nm^3.
8. Tamajit Pain: Taking Stock: India's Metcoal, Coke outlook. Mjunction: innovating for outcomes. https://www.mjunction.in/blog/taking-stock-indias-met-coal-coke-outlook/ (viewed 25/6/2025).
9. https://www.focus-economics.com/commodities/energy/coking-coal/10/2022 (accessed 25 January 2025).
10. According to SteelMint India, coking coal prices in India had moved up by around $100 per tonnes in the last six months to trade at $350 per tonnes at the end of 2023. Published in The Hindu November 06, 2023. https://www.thehindu.

com/business/Industry/india-to-remain-export-market-for-coking-coal-rising-prices-remain-a-concern-industry-body-isa/article67503540.ece

11. Manjeet Kaur et al.: District-wise Cost of Cultivation of Important Crops in Punjab. Department of Economics and Sociology, Punjab Agricultural University, February 2018, Executive Brief.
12. Rural Electrification in India: Customer Behaviour and Demand, February 2019. https://www.rockefellerfoundation.org/report/rural-electrification-india-customer-behaviour-demand/#:~:text=Average%20electricity%20demand%20of%20surveyed,customers%20often%20stacking%20multiple%20sources (viewed 7/10/2022).
13. Rich gas contains 5 or 6 gallons or more of recoverable liquid hydrocarbons per thousand cubic feet, whereas lean gas contains less than 1 gallon of recoverable liquid hydrocarbons per thousand cubic feet. https://www.hindustanpetroleum.com/cng#a3 (viewed 7/10/2022).
14. The Census of India 2011 put it at 27.7 million.
15. The ratio of the final to the initial increase in incomes is called the 'multiplier'. The British economist John Maynard Keynes showed in the 1930s that the final increase in income will be the inverse of the national rate of saving. In India the current rate of saving is 28%, so the multiplier is slightly below 4. However, if most of the increase in income described in the text takes place within the poorer segments of society, as is likely if it is spent locally in the villages, the lower rate of saving of the poor could yield a higher final increase in national income. The concept of the multiplier was introduced in Chapter 14 and was used in a number of previous chapters.
16. Calculated from data collected during the Census of India 2011. See Tables in Appendices A3 and A4.

16

Summing Up: The Path to Net Zero and Human Survival

The last hundred years have been a wondrous era of discovery. At its beginning we had a fair idea of our universe: we knew that it contained an unknown number of galaxies, each of which contained millions of stars, and several of which we surmised might have planetary systems similar to ours. And we wondered whether intelligent life forms might have developed in any one or more of them. We also had a decades' long discussion of whether the universe was static and therefore, in a non-metaphysical way, eternal or whether it was dynamic and expanding.

Thanks to extraordinary advances in science, by the end of the 20th century we had found definitive answers to many of these questions. We know now that the universe contains not only galaxies but also black holes—incredible concentrations of mass and energy in infinitely tiny spaces. We know that, far from being stationary, the universe began with a 'Big Bang', possibly the explosion of a black hole, 13.8 billion years ago, and that the force released by that explosion is still causing it to expand. And we have reason to believe that possibly when it is spent, a few billion years in the future, it may begin to contract.[1]

We have learned, from the carbon dating of fossilized remains and from 'Lucy', the first more or less complete skeleton discovered at the northern end of the Great Rift Valley in 1974, that hominids already inhabited the Earth 3 million years ago. We learned that the ascent of *Homo sapiens* to its present towering eminence began around 150,000 years ago and continued while our early ancestors were driven to migrate by the deterioration of the climate during the Ice Age that dominated the global climate for close to 100,000 years, from 120,000 years before present (BP) up to about 20,000 years BP. By 20,000 years BP, humans had spread to all

Sun, Wind, and Biomass. Prem Shankar Jha and Michael B. McElroy, Oxford University Press.
 DOI: 10.1093/9780198944768.003.0016

of the world's continental land masses with the exception of Antarctica. A temporary reversal in climate during a period known as the Younger Dryas about 11,800 years BP prompted the inhabitants of the region known as the Fertile Crescent to develop the seeds of modern agriculture.

Chapter A2 recounts the fascinating story of the emergence of our human ancestors as a force of nature, leading up to their ability to change the global climate, with uncertain consequences not only for humanity but also for the future of most other forms of life on Earth. Today, while all the findings of modern climate science are warning of the dangers from our continued use of fossil fuels, there is still a marked reluctance on the part of governments to alter current policies to the extent required to avert the possibility of a future catastrophe.

Some 84 countries acknowledged the need to take action a quarter of a century ago by signing the Kyoto Protocol but did next to nothing to honour their commitments. Over this quarter century, the threat of irreversible climate change from global warming grew radically. Up to the 1990s and the beginning of the present century, the bulk of the greenhouse gas emissions responsible for the climate problem originated from the industrialized countries that constitute the OECD. But since then, emissions from developing countries have increasingly emerged as the main driving force for global warming, doubling from 17 billion tonnes a year in 1990 to 34.4 billion tonnes in 2019. This has changed the challenge the world faces from the relatively simple task of replacing one set of energy technologies with another to the more daunting need to do this in a manner that does not impede the alleviation of poverty, especially for the less-advantaged members of the global community.

The world recognized the need for this change of objective in January 2016 by amending its '*Millennial* Development Goals' to '*Sustainable* Development Goals'. These goals required countries to 'end all forms of poverty, fight inequalities *and tackle climate change, while ensuring that no one is left behind*'. This was a brave commitment, but hollow, for it was not accompanied by a suggestion as to how the goal could be achieved. The signatories implicitly placed all their trust on science to find the solutions they would need. Scientists responded by bringing about radical reductions in the cost of solar and wind energy, but these advances have so far made inroads only into cutting back on the use of fossil fuels for power generation. The major portion of the road ahead remains to be charted.

Every government, non-governmental organization, and think tank in the world is aware of this, but the furthest we have gone towards finding solutions is to spell out with ever-growing precision the targets that need to be met and to highlight every new technology that promises to reduce our dependence on fossil fuels in the future. We have not yet begun to identify the challenges that the harnessing of these technologies present for countries in different climatic zones, at different levels of economic development, and with different priorities for economic and social development.

So far, governments in the developing world have taken it for granted that the shift will be difficult. It will require vast sacrifices because an entire industrial infrastructure that has been built painstakingly, only recently, and at great cost will have to be prematurely retired, and savings that would otherwise have gone into building new industry and infrastructure, and creating more jobs, will have to be diverted to replacing the existing infrastructure with a new, more climate-friendly alternative.

It is necessary therefore for all decision makers, be they members of their governments or decision makers in the energy and related industries around the world, to accept that only decarbonization technologies that also yield significant increases in GDP and employment will be readily accepted by the people and governments of the developing world in the limited time we have left.

That is the void that we have tried to fill in this book. We chose to concentrate on India for two reasons. First, it is the third largest and, despite its small per capita carbon footprint, the fastest growing national emitter of greenhouse gases in world. Second, the wide variety of climatic regions within the country make India an ideal location to explore the combinations of technology and policy that can enable the developing world to harness the three primary energy sources upon which it must depend if it is to successfully address the twin challenges of decarbonization and economic development: sun, wind, and biomass. The climatic zones in India range from the cold, scarcely inhabited plains of the trans-Himalayas to the hot, densely settled regions of the Indo-Gangetic Valley, extending to the semi-arid, rain-dependent environments of the centre and south.

In the preceding chapters we presented an analysis of the current state of the Indian economy and society (Chapters 3–6); the government's largely ineffectual attempts to shift out of fossil fuels without harming current production and employment growth (Chapters 7 and 8); the country's enormous, and as yet almost wholly untapped, potential for harnessing sunlight, wind, and biomass to replace coal, oil, and gas (Chapters 9–12); the economic incentives that future Indian governments will need to offer and the policy adjustments they will need to make to gain the confidence of investors, while eliciting cooperation from the owners and customary users of the vast amounts of land that will be needed to facilitate the shift to renewable energy (Chapter 13).

In Chapters 14 and 15 we showed how the solar, wind, and biomass technologies described earlier can help to end the crises in dryland and irrigated agriculture that have strangled growth in rural areas, that have led to farmers committing suicide on an average of one every 35–45 minutes, and that have made it necessary for millions of young men and women to migrate to distant cities in desperate search for work.

Shifting the locus of development from cities to rural areas

We pointed out that since harnessing the effectively inexhaustible energy of the sun, wind, and biomass requires commitments of vast amounts of land, a sharp increase in rural investment will be needed to create a large number of stable jobs on these jobseekers' doorsteps. Their consumption expenditure can raise the incomes of the recipients and trigger further rounds of expenditure and saving in the economy until all of the primary injection of income is absorbed into national savings. The resulting stimulus to demand can trigger additional investment in industry, construction, and the service industries, leading to further increases in GDP and employment, while giving the low-income countries of the developing world precisely the kind of development they need but are not getting with the present emphasis on capital-intensive, urban-centred options for development. In the paragraphs below, we offer a summary of the social and economic benefits that the decarbonization strategy described in this book can bestow on the poor, especially the rural poor, of India.

Potential economic impact of solar and wind power generation on GDP

If India brings half of the 238,000 km^2 of the country's arid and semi-arid areas under solar power generation, it will be able by 2050 to generate more than 3,500 TWh of electricity from solar energy alone. Backed by battery storage and potentially green hydrogen, this may suffice to decarbonize virtually the entire electricity consumption of the country. A mere 5% royalty imposed on electricity bills distributed to the landowners could generate close to Rs 1.5 trillion ($18 billion) in additional income for the owners and customary rights holders of this land, lifting them permanently out of poverty. Similarly, arranging for as little as half of the 120 mha (120,000 km^2) of rain-dependent and fallow farmlands to host 3-MW wind turbines by 2050 could create 360 GW of generating capacity, adding Rs 2.44 trillion ($30 billion) to the income of small and marginal farmers in the eight rain-dependent states that have the maximum wind power potential but also the highest contemporary suicide rates. Spending of this nearly Rs 4 trillion of additional income could increase GDP by up to five times the initial increase in earnings.

Impact on employment

In addition to providing an assured income to the owners of and to the holders of rights to the vast amounts of land that solar and wind power generation will need, these technologies can create a significant number of additional jobs in rural areas. A rough estimate can be inferred from the current ratio of power generated to jobs created. At the end of 2021, India's solar and wind energy industries had installed capacities of 48.5 GW and 40 GW respectively and were employing 111,400 workers.[2] If this labour to output ratio were to remain unchanged, by 2050 ***these industries could create close to 4 million additional permanent jobs in the rural areas of the country.***

Two-stage biomass gasification: the way to harness crop residues

The above benefits are significant, but they could be dwarfed by the income and jobs that could flow from the harnessing of biomass energy through the gasification of agro-industrial residues, municipal solid waste, and crop residues, which we described in Chapter 12. As we indicated there, agro-industrial and municipal residues together could yield up to 800 Mt of methanol, or equivalent amounts of other liquid fuels.[3] But that is only a little over half of the potential of biomass that could be deployed to meet India's future demand for transport fuels without disturbing the existing pattern of agriculture. The single largest biomass source is the residue from India's cereal crops. Several estimates of the magnitude of this source have been made over the years. In 2018, the Government of India's Technology Information Forecasting and Assessment Council (TIFAC) and the Indian Agricultural Research Institute (IARI) carried out a joint survey of the total availability of crop residues that could be deployed for conversion into biofuels in the 662 districts[4] of the country. The data that formed the basis of this survey were the production statistics from 2010–11 to 2015–16 for eleven crops, namely rice, wheat, maize, sugar cane, cotton, pulses, and oilseeds (groundnut, mustard, soybean, and castor). The selection of these crops was based on an earlier study of their acreage and total production across the country.

Averaging the yields over those 5 years, the study estimated that the total *dry* weight of crop residues generated by these eleven crops amounted to 683 Mt. Eighty percent of this came from just four crops: rice straw and husk (228 Mt); wheat straw (150 Mt); sugar cane tops and bagasse (116 Mt); and cotton (54 Mt). The remaining 136 Mt came from maize, lentils, grain, and five oilseeds—soyabean, rapeseed, mustard, groundnut, and castor. Excluding sugar cane residues, whose use has been discussed earlier, this amounts to 567 million dry tonnes or 708 Mt with a moisture content of 25%, the optimum for biomass gasification. As noted in Chapters 12 and 15, these crop residues cannot be fed directly into modern, oxygen-blown and hydrogen-added gasifiers because these facilities need to be large in order to be economical. They would therefore require huge amounts of crop residues to be transported over long distances to meet their demand. But as we pointed out in Chapter 15, this problem could be resolved if the crop residues were first gasified in simple air-blown gasifiers in the villages.

Gasification of the crop residues in the villages could provide a significant source of lean gas to fuel cold storage units together with 585 Mt of methanol-equivalent biofuels. In sum, the readily available biomass in India could supply close to 1,400 Mt of methanol-equivalent transport fuels while meeting a significant part of the need for electricity where this is most urgent. In terms of energy, this would be equivalent to more than four times the energy consumed in the form of oil and natural gas in the transportation sector today.[5] Further, in contrast to the limits imposed by the land available for solar and wind power, the future supply of biomass is elastic. According to the Forest Survey of India, 24.2% of the land surface is covered by 'forests and tree cover'.[6] But this estimate is deceptive, given that a major portion of the forest cover has been illegally felled by timber smugglers who have left the land either entirely denuded or covered by sparse secondary growth. A guaranteed and ever-increasing demand for biomass could galvanize the state forest departments into pursuing an aggressive policy of reafforestation. Not only could this supply a major fraction of the biomass that will be needed in the future but also it could at the same time increase the size of India's natural carbon sink.

Finally, crop residue gasification can go a long way towards clearing the pall of smoke and other aerosols that blankets North India every winter. In Chapter 12, we pointed out that the use of agricultural crop residues to produce modern transport fuels has so far been foiled by the exorbitant cost of transporting these resources to facilities for their processing. So, apart from marginal demands for a few commercial uses such as specialty papermaking, the rest has been put to non-market uses in the villages, such as digestion in biogas plants to produce cooking gas, thatching roofs, and bedding down milch cows and buffaloes housed in cowsheds. As was pointed out in Chapters 5 and 13, nearly all of the rice straw and an increasing proportion of the wheat straw is burnt to clear the fields for the next crop. This is the main ingredient for the pall of polluted air that now covers most of north India in November and December every winter. The first-stage gasification of these crop residues in the villages could eliminate this source of air pollution.

Employment-generating potential of biomass

In Chapter 15, we described how a gasifier–cold store combination in every village could generate 50 or more permanent jobs in each village and, therefore, over 600,000 permanent jobs in the 12,278 villages of the state of Punjab. India has 640,000 villages, of which 100,000 or thereabouts are very small and will need to be served by their larger neighbours.[7] Another 93,000 villages have grown into small townships and now house a large part of the country's small scale industry. In the rest of the country, the average yield per hectare of the two primary cereal crops is a little over half that of Punjab and for vegetables about two-thirds. Assuming that their needs can be met by 400,000 gasifiers and cold storages similar to those employed in Punjab, this could lead to a primary generation of as many as 20 million jobs. When the multiplier effect created by the resulting increase in spending is fully absorbed by the economy, this number could rise to 80 million or more. Assuming this takes three decades, it will mean the addition of an average of more than 4 million jobs a year in the rural areas and adjoining small towns.

The multiplier effect of import reduction on national income and employment

Although important, the income and employment effect described above still do not account for the full impact that harnessing biomass to energy generation could have on India's GDP, employment, and the demand for labour. Prior to the COVID-19 pandemic and the Ukraine War, India was importing 210 Mt of oil per year, at a cost of $110 billion or thereabouts.[8] Since this money was going out of the country, the benefit from the multiplier effect of this spending was going to the oil-exporting countries and the countries from which they purchased their needs. Self-sufficiency in transport fuels will release this money into India's domestic economy and add more than $400 billion to India's GDP. That will increase the country's GDP by up to 15% of its 2020–21 level and have a corresponding effect on employment. Over a 30-year period this will increase the rate of growth of GDP by at least half a percent a year. The additional employment this could generate is for econometricians to estimate.

Notes and References

1. Andrei, Cosmin, Ijjas, Anna, and Steinhardt, Paul J. (2022). Rapidly descending dark energy and the end of cosmic expansion. *PNAS* 119(15):e2200539119. https://www.pnas.org/doi/full/10.1073/pnas.2200539119 (viewed 9/11/2022).
2. Charlotte Steiner and Sameer Kwatra (co-authored with Charu Lata and Akanksha Tyagi): India Could Create Millions of Jobs Through Renewable Energy. Natural Resources Defence Council, 27 January 2022. https://www.nrdc.org/experts/sameer-kwatra/india-could-create-millions-jobs-through-renewable-energy (viewed 9/11/2022).
3. MIT Researchers: Methanol is a viable alternative to gasoline. Fuel Freedom Foundation, 22 October 2015. Leslie Bromberg and Daniel R. Cohn of the MIT Energy Initiative spent more than a year exploring the advantages and disadvantages of using methanol in gasoline engines. They concluded that methanol offers both high flame speed and high octane when compared to gasoline. 'These properties can increase the efficiency by 20–30% relative to standard port fuel injected gasoline engine by optimized use of existing engine components.' https://www.fuelfreedom.org/mit-researchers-methanol-is-a-viable-transportation-fuel/ (viewed 10/11/2022).
4. The basic unit of administration in the states of the Indian Union.
5. Gasoline contains 2.2 times the chemical energy of methanol, but methanol's flame speed is around 1.3 times that of gasoline. The consumption of liquid transport fuels in 2019–20 required 165 Mt of crude oil.
6. Ministry of Environment, Forest and Climate Change: India State of Forest Report 2021, released 22 January 2022. chrome-extension://efaidnbmnnnibpcajpcglclefindmkaj/https://static.pib.gov.in/WriteReadData/specificdocs/documents/2022/jan/doc20221207001.pdf (viewed 15/11/2022).
7. Census of India 2011.
8. India's oil import bill doubles to $119 billion in FY22. *The Economic Times*, 24 April 2022. https://economictimes.indiatimes.com/industry/energy/oil-gas/indias-oil-import-bill-doubles-to-usd-119-bn-in-fy22/articleshow/91049349.cms (viewed 20/11/2022).

ANNEXURE 1

Earth: A Dynamic Planet, Fuelled by the Sun, Shaped by Life and by Tectonic Forces that Determine the Properties of its Surface

The energy budget of the Earth's climate is regulated primarily by input from the sun, balanced by compensating emission to space. The solar source is delivered mainly in the form of visible light originating from a region of the solar atmosphere known as the photosphere defined by a temperature of about 5,600 °C. The higher the temperature of an object, the shorter the wavelength of the light it emits. The solar energy is emitted primarily in the visible. The energy emitted is a function also of temperature, varying under ideal circumstances as the fourth power of the temperature. The energy emitted by the sun amounts to 63 million watts per square metre of surface area (63 million W/m^2). By the time it reaches the orbit of the Earth, the solar source has been diluted to 1,370 W/m^2).

Earth at its orbital distance from the sun represents a relatively small target to capture the solar energy as it diffuses through space. A portion of the solar energy intercepted by the Earth is reflected back to space, primarily by clouds in the atmosphere. The fraction reflected is referred to as the *albedo*. The current albedo of the Earth amounts to 31.3%: so, 68.7% of the incident energy is absorbed, mainly by dark surfaces, exemplified, for example, by forests and the ocean.

The energy absorbed by the Earth must be compensated by energy emitted to space. To balance the energy absorbed, the Earth needs to radiate energy at an effective temperature of about −18 °C. Clearly this temperature cannot be identified with average conditions at the planetary surface. It refers rather to a region of the atmosphere at an elevation of about 5 km above the surface. The reason that the surface is unable to communicate heat directly to space has to do with the presence of trace gases in the atmosphere that are effective in trapping a significant fraction of the planet's infrared heat emission. The most important of these heat-absorbing gases is water vapour. But the abundance of water vapour in the atmosphere is determined primarily by the temperature at the surface. This is where the presence of the other absorbing gases becomes important, notably infrared trapping gases such as carbon dioxide (CO_2) and methane (CH_4) (the nitrogen and oxygen molecules that compose the bulk of the atmosphere are ineffective in absorbing infrared radiation). If these heat-absorbing gases were not present, the ocean would be frozen over, the planet would be freezing cold, and the supply of water vapour to the atmosphere would be minimal. These heat-absorbing compounds are referred to as *greenhouse gases* based on an erroneous interpretation of why the air inside a greenhouse is warmer than the exterior. The air is not warmer in the inside because the heat is trapped by the glass but rather because the glass, while allowing sunlight to penetrate, inhibits the warm air in the interior from communicating directly with the outside.

Even a small imbalance in the Earth's energy budget would have important implications for the planet's climate: an excess in emission could cause the Earth to freeze over; conversely, a deficit and the planet would undergo significant warming. The latter situation is what we are confronting today. The global average surface temperature has increased by more than 1 °C over the past century and promises to increase by much more over the immediate future if we are unable to arrest the factors responsible for the warming, notably the increase in emissions of CO_2 associated with society's reliance on fossil fuels—coal, oil, and natural gas—as the primary source of energy for our modern industrial economy, compounded by emissions of CH_4 from a variety of sources including notably agriculture, specifically animal husbandry and rice cultivation. A key objective of the current study is to explore strategies to reduce or even reverse the current trend, with a particular focus on India.

The ocean plays an important role in regulating climate, specifically as a reservoir to store heat and to redistribute this heat. Think of the function of the Gulf Stream in transferring heat from low to high latitudes in the Atlantic. It is the heat delivered by the Gulf Stream that is largely responsible for the relatively mild climates experienced by countries in north-western Europe, countries such as Ireland, Scotland, and Norway for example. Northern Scotland is at about the same latitude as central Hudson Bay or Alaska. Why the difference in climate? It can be attributed largely to the fact that surface waters in the North Atlantic are saltier relative to waters in the North Pacific. When waters in the high Northern Atlantic cool in winter they become dense enough to sink, setting up a global circulation pattern that has an influence on regions far removed from the original origin of this sinking water. The late Wallace Broecker referred to this phenomenon as the Great Ocean Conveyor Belt.[1] Properties of the Conveyor Belt are illustrated schematically in Figure A1.1.

The Atlantic is a relatively minor ocean compared with the Pacific. Why should the Atlantic play such a major role in regulating the global flow of waters in the deep ocean? Why should the salt content of surface waters in the North Atlantic be greater than the corresponding salt content of waters in the North Pacific? The underlying reason for the buildup of salt in the Atlantic is the presence of the Rocky Mountain range running virtually the entire length of the west coast of North America. This restricts the transfer of water vapour from the Pacific to the Atlantic. By contrast, there is no similar impediment to the transfer of vapour from the Atlantic to the Pacific over the Eurasian land mass, reflecting the fact that the Alps and the Himalayas run not north to south but west to east. The asymmetry of the topographical configuration between the two great land masses of the Earth is ultimately responsible for the circulation of the global oceans reflected in Broecker's Global Conveyor Belt.

As indicated schematically in Figure A1.1, deep water formed in the North Atlantic flows south in the Atlantic, joined by a second source of deep water contributed by cold salty waters off the coast of Antarctica. The two sources are believed to be comparable in magnitude. A portion of the deep water is diverted into the Indian Ocean, with an additional source delivered to the Pacific. The deep water percolates slowly to the surface as it progresses on its lengthy journey around the planet. The resulting shallow water makes the return loop to the North Atlantic, replacing the deep water

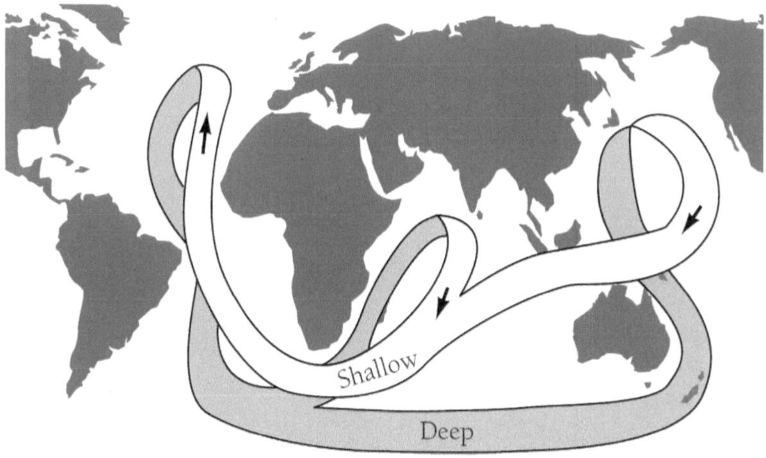

Figure A1.1 The Conveyer Belt.
Source: McElroy, Michael B. *Energy: Perspectives, Problems, and Prospects*. OUP, USA, 2010, based on Langmuir, Charles H., and Wally Broecker. *How to Build a Habitable Planet*. Princeton University Press, 2012.

that formed originally in that region. The circulation is fuelled by an impressive volume of water, estimated by Langmuir and Broecker[2] to be as much as 100 times the volume of water delivered to the ocean by the world's largest river, the Amazon.

There were times in the past when the Earth was much warmer than it is today and times when it was much colder. Tectonic processes played an important role in regulating this long-term climate variability. The influence was two-fold. Movements of the crustal plates were responsible for changes in the configuration of the continents, with implications both for the spatial absorption of sunlight and for the ability of the ocean to redistribute this energy. Even more important perhaps was the role tectonic processes could play in altering the abundance of CO_2 in the atmosphere. The bulk of the Earth's carbon resides in the sediments, formed by deposition at the bottom of the world's oceans. Tectonic processes are responsible for recycling this material. Weathering of continental rocks provides a sink for atmospheric CO_2. Conversely, CO_2 is returned to the atmosphere as a component of volcanos and hot springs, as carbon-rich sedimentary material is drawn down into the mantle in response to the convergence of crustal plates.

This cycling need not proceed at a constant pace over geologic time. There were times when it could have contributed to an increase of CO_2 in the atmosphere, and times when it could have been responsible for a deficit. Climate would have adjusted accordingly.

The geological record documents periods when the climate was unusually warm—temperatures were moderate even at high latitude in winter. It records also times when the planet was freezing cold even at the equator, a condition referred to by California Institute of Technology geologist Joseph Kirschvink as Snowball Earth.[3] The changing configuration of the continents responsible for these disparate influences over time is illustrated in Figure A1.2.

510 Ma (Paleozoic/Ordovician/Canadian/Tremadoc)

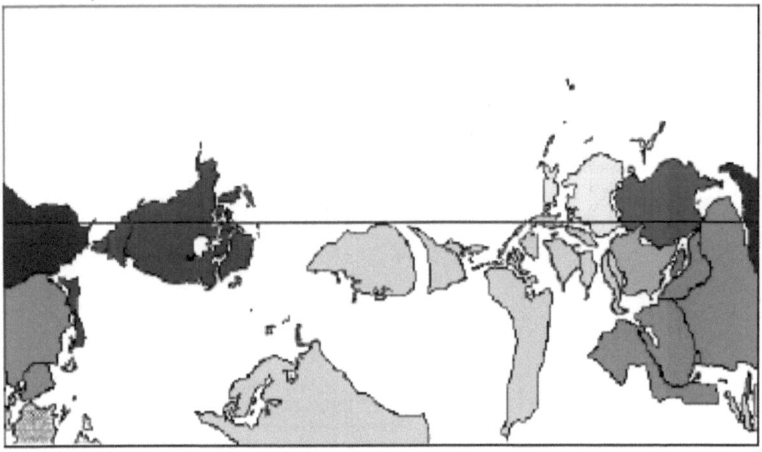

410 Ma (Paleozoic/Silurian/Pridoli)

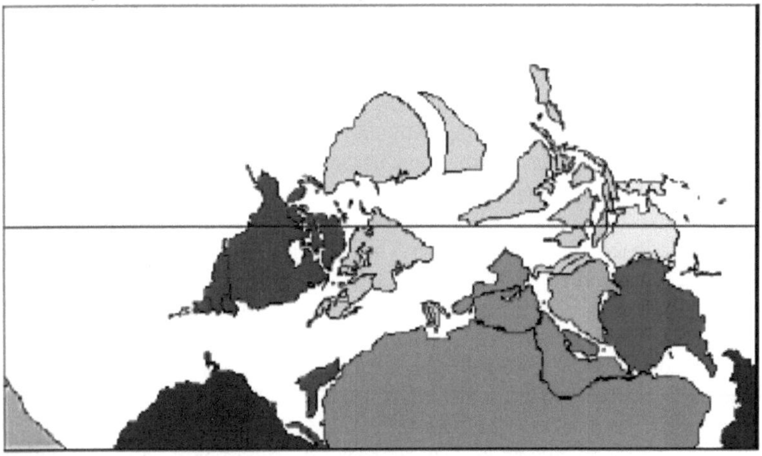

Figure A1.2 The changing configuration of continents.

Endnote 3: 50 Ma means 50 million years ago. Cenozoic is the current geological era, spanning from about 66 million years ago to the present. Tertiarys is a term that describes the period from about 66 million to 2.6 million years ago. Paleogene: This is a period within the Cenozoic era, from about 66 million to 23 million years ago. Eocene: This is an epoch within the Paleogene period, spanning from about 56 to 34 million years ago. Lutetian is a stage within the Eocene epoch, ranging from about 48.6 to 41.3 million years ago.

Source: McElroy, Michael B. *Energy: Perspectives, Problems, and Prospects*. OUP, USA, 2010, based on Windley, B.F., *The Evolving Continents*. 3rd Edition, John Wiley and Sons, New York, 1995.

320 Ma (Paleozoic/Carboniferous/Pennsylvanian/Bashkirian/Yeadonian)

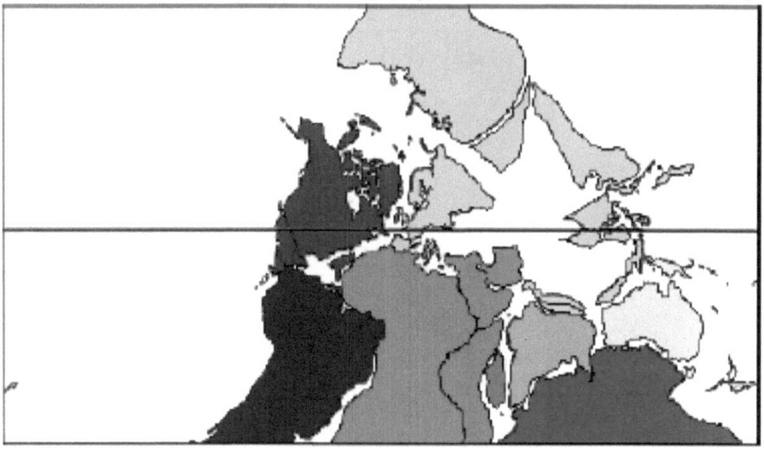

260 Ma (Paleozoic/Permian/Rotliegendes/Artinskian)

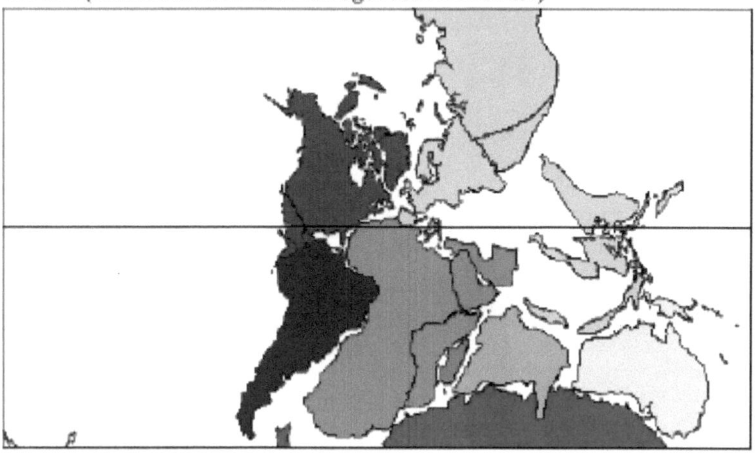

Figure A1.2 Continued

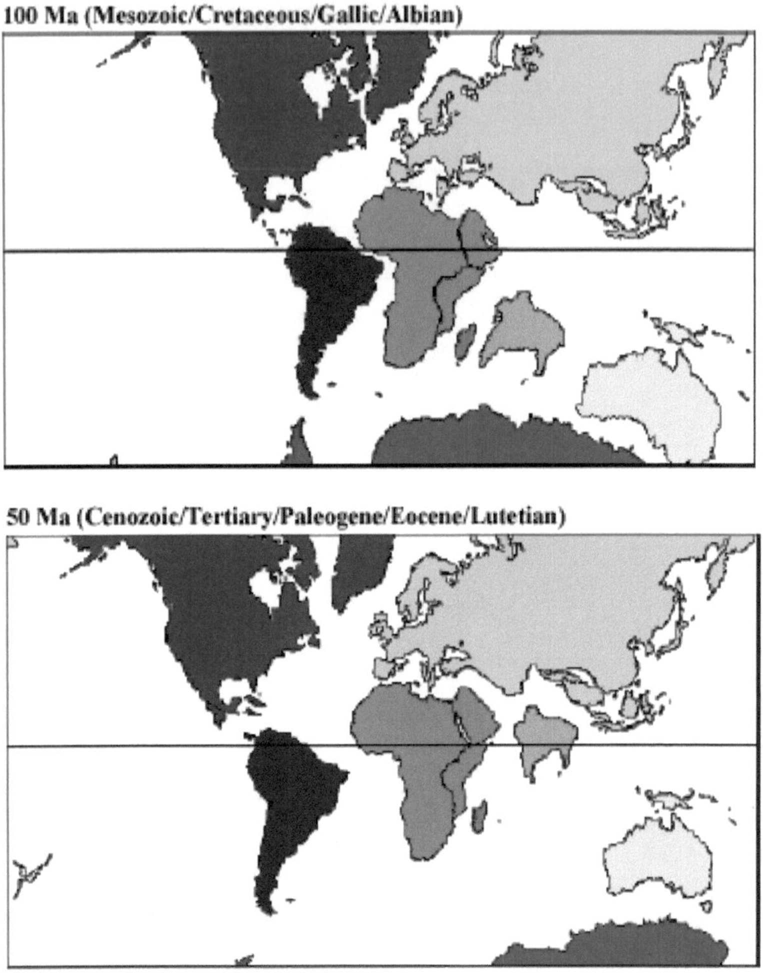

Figure A1.2 Continued

The changes in CO_2 that resulted from the forces described above, though often very great, were natural and occurred over long periods of time. Today we are returning CO_2 from sediments to the atmosphere at a rate 50 times faster than nature, by mining and burning the fossil fuels that have provided the backbone of our global energy system over the past several centuries. In terms of the carbon cycle, we are functioning inadvertently as an amplifier of the world's volcanos and hot springs. It is not surprising therefore that the concentration of CO_2 is on an upward trajectory not only in the atmosphere but also in the ocean.

Life has been a persistent presence on Earth from an early stage of the planet's history, for at least 3.5 billion years and arguably longer. Photosynthesis, the combination of CO_2 and H_2O, fuelled by light to form organic matter, represents the fundamental first step in the synthesis of living organisms. This process is associated with the production of oxygen. The earliest forms of life consisted of organisms known as prokaryotes.

Bacteria and blue-green algae are examples of these simple life forms that developed over the first few billion years of Earth's history. Bacteria function as nature's garbage disposers, transforming all kinds of organic waste into more useful compounds. Blue-green algae, for example, have the capacity to use sunlight to convert chemically inert nitrogen from the atmosphere into more useful forms such as ammonia that can be readily incorporated in the structural material of plants and animals. To accomplish the same task today, we rely on chemical factories consuming vast quantities of fossil fuels, although, as we will see later, we are now developing the ability to produce this fertilizer, what we refer to as green ammonia, using hydrogen extracted from water by a process referred to as electrolysis, fuelled by a carbon-free source of electricity developed, for example, from either wind or PV solar panels. Bacteria play an additionally important role in converting low-grade organic matter to higher quality organic material to feed ruminants, such as cattle, sheep, and goats. The downside in this case is the production of climate-altering methane (CH_4).

The next stage of evolution involved the appearance of eukaryotes, unicellular organisms significantly more complex than their prokaryotic predecessors. This development occurred approximately 1.5 billion years ago. The evolution of the eukaryotic cell set the stage for the formation of more complex multicellular organisms, eventually for plants and animals, and ultimately humans. An important intermediate step was the development of an oxygen-mediating enzyme that allowed eukaryotic organisms to survive in the presence of free oxygen, paving the way for the buildup of oxygen in the atmosphere and for the synthesis of ozone that could protect life at the surface from otherwise lethal doses of DNA-altering ultraviolet solar radiation. This monumental event occurred about 440 million years ago, signalling a new era of planetary biological evolution identified with episodes when large numbers of new species appeared over relatively brief periods of geologic time, offset by times when important numbers of existing species became extinct, responding either to changing environmental conditions or by losing out to better-adapted competition.

The explosion of new life forms that arose during the Cambrian Period [543 million to 510 million years before present (BP)] was followed a few million years later by what the late Stephen Gould[4] referred to as the 'granddaddy of all extinctions' responsible for elimination of as much as 95% of all marine species alive at that time. A second major extinction, triggered in this case by the impact of a giant meteorite, resulted in the demise of the dinosaurs, paving the way for the later ascendancy of large mammals and eventually our human ancestors. As Gould pointed out, mammals spent the first million years of their existence 'as small creatures living in the nooks and crannies of a dinosaur's world'. The demise of the dinosaurs was a happy event for the mammalian world and indeed for humans. Gould drolly summed the situation up thus: 'in an entirely literal sense, we owe our existence, as large and reasoning animals, to our lucky stars'.

Table A1.1 Chronology of important events in Earth's history.

Earth events	Time (years BP)	One-year perspective
Earth forms	4.50E + 09	1 January
Eukaryotes	1.50E + 09	1 September
Burgess Shale	5.00E + 08	20 November
Vascular plants	4.45E + 08	24 November, 9pm
Life expands to land	4.40E + 08	25 November, 7am
Amphibians	3.00E + 08	6 December, 4pm
Massive extinction	2.25E + 08	12 December, 6pm
Mammals	1.60E + 08	18 December,
Dinosaurs eliminated	6.50E + 07	25 December, 5pm
Homo erectus	1.70E + 06	31 December, 8pm, 41min
Homo sapiens	1.50E + 05	31 December, 11pm, 42min, 29sec
Human migration	5.00E + 04	31 December, 11pm, 54min, 10sec
Humans reach Americas	3.00E + 04	31 December, 11pm, 56min, 30sec
Agriculture	1.00E + 04	31 December, 11pm, 58min, 50sec
Industrial Revolution	250	31 December, 11pm, 59min, 58sec

It is difficult to keep track of the momentous events of Earth history when the story is recorded in terms of large intervals of time, in millions or even billions of years BP. To provide a more intuitive perspective, we find it useful to summarize the important milestones not only in terms of real years BP but also by compressing the 4.5-billion-year history of the planet into a single illustrative hypothetical year. The results of this exercise are presented in the accompanying Table A1.1. Note the late arrival of our human antecedents—deep in the evening of 31 December.

Only in the last few seconds of this metaphorical year have humans developed the capacity to alter the environment of the Earth on a global scale. The challenge, as posed by McElroy[5], is 'whether we can develop the wisdom and foresight to ensure a harmonious co-existence of our species with nature over the first few seconds of the year that is about to begin'.

Following the more general introduction in this chapter to the history of climate and life on Earth, we turn our attention now in Chapter A2 to a more focused discussion of the emergence of humans as a force for global change.

Notes and References

1. Langmuir, Charles H. and Broecker, Wally (2012). *How to Build a Habitable Planet*. Princeton University Press.
2. Ibid.

3. Kirschvink, Joseph L. (1992). Late Proterozoic low-latitude global glaciation: the Snowball Earth. In: Schopf, J. and Klein, C. (eds), *The Proterozoic Biosphere: A Multidisciplinary Study.* Cambridge University Press, pp. 51–2.
4. Gould, Stephen Jay (1990). *The Burgess Shale and the Nature of History.* Hutchinson Radius.
5. McElroy, Michael B. (2010). *Energy: Perspectives, Problems, and Prospects.* Oxford University Press.

ANNEXURE 2

Emergence of Humans as a Force for Global Change

Our closest relatives in the animal kingdom are the giant apes (including the gorilla and the chimpanzee). Much of the development that led to the appearance of modern humans is believed to have taken place in Africa. The evolutionary sequence is, however, unclear. It is believed to have proceeded along a trajectory that sequentially involved the appearance of *Australopithecus africanus*, *Homo habilis*, and *Homo erectus*. *Homo erectus* arrived on the scene about 1.7 million years ago and evolved later into *Homo sapiens*. Interpretations of human mitochondrial DNA suggest that we share a common maternal ancestor, a primitive Eve, and that she may have lived in Africa about 150,000 years ago.

As Jared Diamond[1] tells the story, our early ancestors were nomadic—hunter-gatherers—constantly on the move in search of new plant and animal species that could provide sources of food. Changes in climate forced them to widen their search continuously: for instance, thousand-year droughts, evidence for which has been found under the bed of the Dead Sea, might have driven them to migrate to the seashore, to learn to fish and crack open shellfish, and to thus add fish and crustaceans to their diet.[2] The long periods of drought may also have determined the pattern of their migration.

The other major determinant of human migration was the onset of the Great Ice Age. The Earth was relatively warm 120,000 years ago, so food and fibre would have been readily available in tropical and subtropical Africa. But the climate deteriorated as the world entered the last Great Ice Age. This persisted, with intermittent warm periods, for close to 100,000 years and ended only about 20,000 years ago.

As the Ice Age deepened, it would probably have driven the early humans to migrate from the highlands of central and eastern Africa north towards the Mediterranean, east along the coast to Asia and south towards the warmer coastal regions of Southern Africa. In the warmer interregnums of the Ice Age, they would have pushed further north into China and populated the temperate regions of Europe and Asia. They had reached New Guinea and Australia by 40,000 years BP and had made their way to Siberia and eventually to the Americas by the time the Ice Age had reached its peak 20,000 years ago. Sea level at that time was about 120 m lower than it is today. Asia was connected to the Americas by way of a land bridge across what is now the Bering Strait, providing a convenient passage for transit. Within a few thousand years, migrating largely down the ice-free west coast, humans had expanded their presence to encompass most of the habitable portions of North and South America (the exception, the region occupied by the receding ice sheet that would cover most of Canada for a further 6,000 or 7,000 years). By 15,000 years BP, humans had extended their range to include most of the world's continental landmasses with the exception of Antarctica.

The total population of humans in the hunter-gatherer era was relatively small, a few million or less.[3] To put this in context, there were probably more bison (buffalo) roaming the plains and grasslands of the Americas a few tens of thousands of years ago than there were humans on the Earth as a whole: the population of bison when the Europeans first arrived in the Americas has been estimated at between 20 and 30 million years ago.[4] Despite their small numbers, these early humans had an impressive, though largely negative, impact on the environment. For instance, most of the native large mammals of Australia disappeared a short time after humans arrived. A similar fate befell the bulk of the megafauna in the Americas.

Diamond[5] suggests that 15,000 years ago the American West probably looked much like the Serengeti Plain does today in Africa, with 'herds of elephants and horses pursued by lions and cheetahs and joined by members of such exotic species as camels and giant ground sloths.' After human beings arrived on the scene, this spectacular diversity ceased to exist in a blink of geological time.

There can be little doubt that humans were responsible for this collapse. Diamond[6] suggests that the large animals that populated Africa and Eurasia had survived because they had evolved side by side with humans for a long time. So their capacity to evade human hunters had kept pace more or less with the development of humans' hunting skills. In sharp contrast, the large mammals that populated Australia and the Americas had no such experience and had little chance to survive their first encounter with our environmentally insensitive, voracious ancestors.

We have only recently begun to appreciate the extent to which our future depends on a harmonious coexistence with nature. The late E.O. Wilson, one of the greatest natural scientists of our time, and a pioneer in efforts to protect the *biodiversity* of our planet argued, that we are in the midst today of an extinction event that rivals any the world has witnessed over at least the past 500 million years, greater even than the momentous event that terminated the era of the dinosaurs. The weapons we employ now are not the spears or bows and arrows of our hunter-gatherer ancestors but the fire and bulldozers we use to transform landscapes and to eliminate whole ecosystems in the quest for profit. Nowhere is this problem more serious than in the tropics. where the diversity of life is greatest and where community structures are most complex. A recent study by the United Nations suggests that 25% of the species living in tropical forests today may be doomed to extinction over the next few decades if current trends are not reversed. Deforestation in the tropics is responsible also for significant emissions of CO_2 contributing to the problem of global climate change (this source is estimated to amount to as much as 30% of the contribution from global combustion of fossil fuels).

Humans found a particularly favourable environment when they arrived in the region known as the Fertile Crescent extending from the northern part of the Red Sea through portions of modern Jordan, Israel, Lebanon, Syria, Turkey, Iraq, and Iran (Figure A2.1). When they reached this region approximately 14,000 years ago, they would have had the sense that they were entering *Nirvana*, a virtual Garden of Eden, rich in edible plants and animals. There were sheep, goats, and the ancestors of modern cattle that could be exploited for meat and milk, for skins and wool, and for bones that could be used to make primitive tools. Living would have been easy.

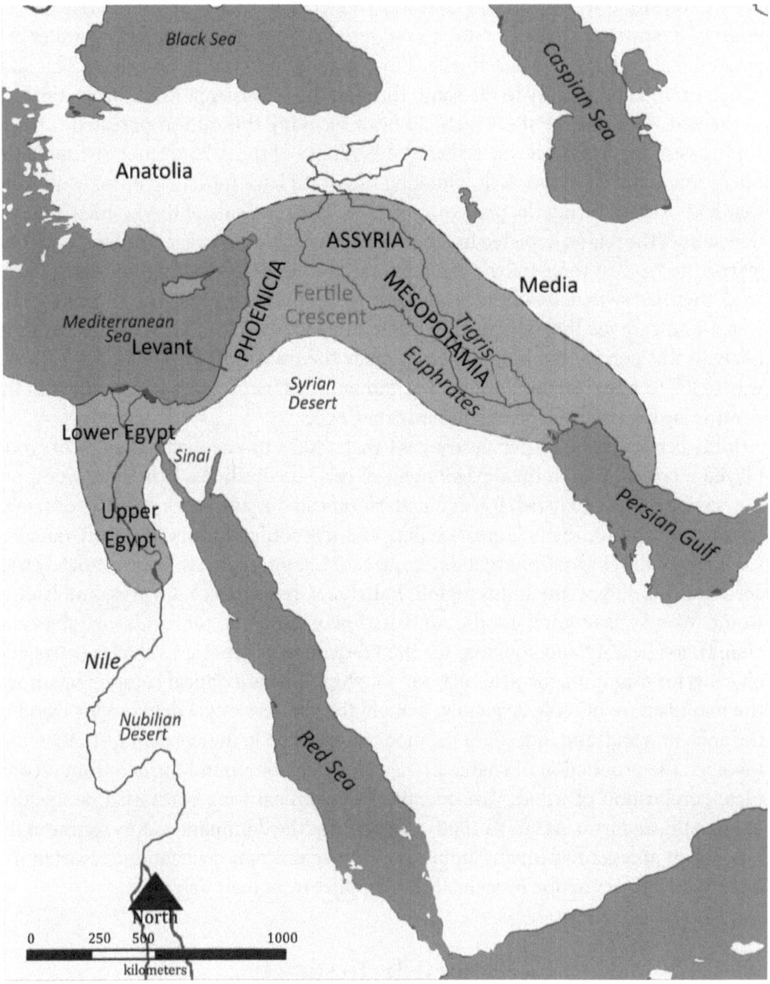

Figure A2.1 The Fertile Crescent in the Middle East.

Source: https://commons.wikimedia.org/wiki/File:Map_of_fertile_crescent.svg (Nafsadh, 22 May 2011).

Recovery of climate from the extremes of the ice age had reached a relatively favourable condition. It is likely that many who arrived in the region during this propitious time would have chosen to settle. Earlier travellers might have found the area less hospitable and would have elected most likely to keep moving. But this benign climate would end abruptly at about 11,800 years BP and the setback would last for close to a thousand years, over a period referred to as the Younger Dryas. During this interval the Earth would plunge back into cold, near-glacial conditions.

The transition from warm to cold was triggered it is thought by the release of a flood of fresh water to the Atlantic through the St Lawrence Seaway. This water originated from a large neoglacial lake, subsequently named Lake Agassiz.[7] The lake, which drained previously to the south through the Mississippi River, broke through to the east when the ice sheet that had been blocking this option retreated. The resulting vast flood of fresh water altered the salinity of the Atlantic and disrupted the operation of the Conveyor Belt, plunging the world back into the climate conditions that had existed during the preceding ice age. The residents of the Fertile Crescent, denizens of the region depicted in Figure A2.1, were able, however, to adapt. They had learned to harvest their water resources, and, rather than foraging for plants in the wild, they had switched to cultivating the varieties they favoured. They began to feed, water, and care for the animals they had come to depend on. In short, it was most likely in this period that human beings made the transition from a hunter-gatherer way of life to one sustained by farming and animal husbandry, a key moment in the continuing saga of the evolving human experience.

John Perlin[8] makes a persuasive case that access to adequate supplies of wood played a critical role in the development of early civilizations. When the wood ran out, civilizations collapsed. It was a pattern repeated many times over the course of history, not just for the first great civilization of recorded history, the Sumerian, but for many of the civilizations that developed contemporaneously or immediately succeeded it—in Egypt, the Indus region, Babylon, Crete, Greece, Cyprus, and later in Rome. Wood was required for the construction of buildings, for furniture (tables and chairs), for heating and cooking, for the fabrication of wheeled vehicles (carts and chariots for example), for weapons, and for ships. It was a critical component also in the manufacture of tools. Arguably, though, the greatest single demand for wood in the ancient world, and indeed for the modern world up to the age of industrialization, involved the production of charcoal, a carbon-rich compound formed from incomplete combustion of wood. Vast quantities of charcoal were consumed in smelting copper, tin, and iron. As late as 1900 CE, wood was the dominant energy source in the US. When access to adequate supplies of timber ran out, civilizations resorted frequently to military action to secure these supplies from their neighbours.

From tribe to society

The first great human civilization developed around 3,500 BCE in Sumeria, on the flood plains of the Tigris and Euphrates Rivers, taking advantage of the rich alluvial soils and perennial supply of water available in that region. To ensure a reliable supply of the latter and to facilitate waterborne transport of supplies, the Sumerians developed an elaborate system of canals.[9] Construction and management of these canals required a complex social structure. Through time, responsibilities became increasingly specialized and distributed: farmers farmed, artisans fabricated tools and constructed drainage ditches, bureaucrats managed the supply and distribution of waters in the canals, merchants engaged in trade, soldiers fought to defend property or to steal from others, and the priests interceded with the gods. The result was the evolution of a hierarchical structure, a pattern that would become the norm for most, if not all, subsequent organized societies.

Trade and conquest played an important role in the success of Sumerian society. Critical resources, other than rich alluvial soil, were in short supply in the homeland. The Sumerians reached out, south through the Persian Gulf and north following the course of the Tigris and Euphrates, to meet their demand for resources such as timber, stone, and metals (notably copper). There can be little doubt that the outreach of the Sumerians for trade would have had a significant impact on the other great civilizations that developed at about the same time in the Valley of the Nile in Egypt and in the Valley of the Indus River in north-western India. Vigorous trade developed between all three regions. Skills once learned are readily transferred.

It is instructive to consider how humans arrived at the dominant role they occupy today. Much of the preceding text has been devoted to describing the history of this odyssey. This is a book, however, about energy and climate. We propose to tell the story now through the lens of energy.

Our hunter-gatherer ancestors would have relied largely on human muscle power. Plant material, first wood and then charcoal, would have been employed primarily for cooking and for staying warm when the weather was uncomfortably cold. The first major advance in our ancestors' command of the environment was their success in harnessing animal power for the purpose of mobility and transport. This occurred at about 3,500 years BCE in Sumeria and was made possible by the invention of the wheel. A few hundred years later, this innovation would lead to the introduction of two-wheeled chariots and their use as powerful weapons of war.

The second major energy shift, from animal to wind and water sources of power, took place in the Babylonian Empire, where Emperor Hammurabi developed ambitious plans to lift water for irrigation and at the same time pioneered the use of windmills. Its systematic deployment, however, dates to the first millennium BCE when the Persians began to implement wind- and water-driven irrigation systems to transfer water over hundreds of kilometres using underground canals known as qanats. Water was transferred from the mountains of Lebanon to the Euphrates River Basin to ensure the steady supply of the water required to produce the food needed to sustain the expanding Persian Empire. At its height, the Persian (followed by the Parthian) Empire stretched from the western edge of modern Greece to the foothills of the Hindu Kush mountains in India and would last for almost 800 years.

By 300 BCE, windmills and water wheels were providing motive power for irrigation, maritime travel, and trade across the Middle East and the Indian subcontinent. A thousand years later, wind power would be harnessed in the fertile lowlands around the North Sea and along the southern Baltic coast for a host of other purposes: for grinding cereals, for lifting water for irrigation, and, significantly, for machining agricultural implements. The resulting increase in productivity led to huge surpluses that could be traded for other commodities. That led to specialization and gave birth to market towns where traders and producers met to exchange their wares. Within little more than a century these market towns coalesced into the Hanseatic League with close to 200 cities. The League went on to dominate trade in Europe from the 13th to the 15th centuries CE. The region that was occupied by the Hanseatic League is illustrated in Figure A2.2.

By far the most important development in the energy saga must be attributed, however, to the transition that led to the use of fossil fuels—coal, initially, and subsequently oil and natural gas. Our forebears in this case were tapping a carbon-based

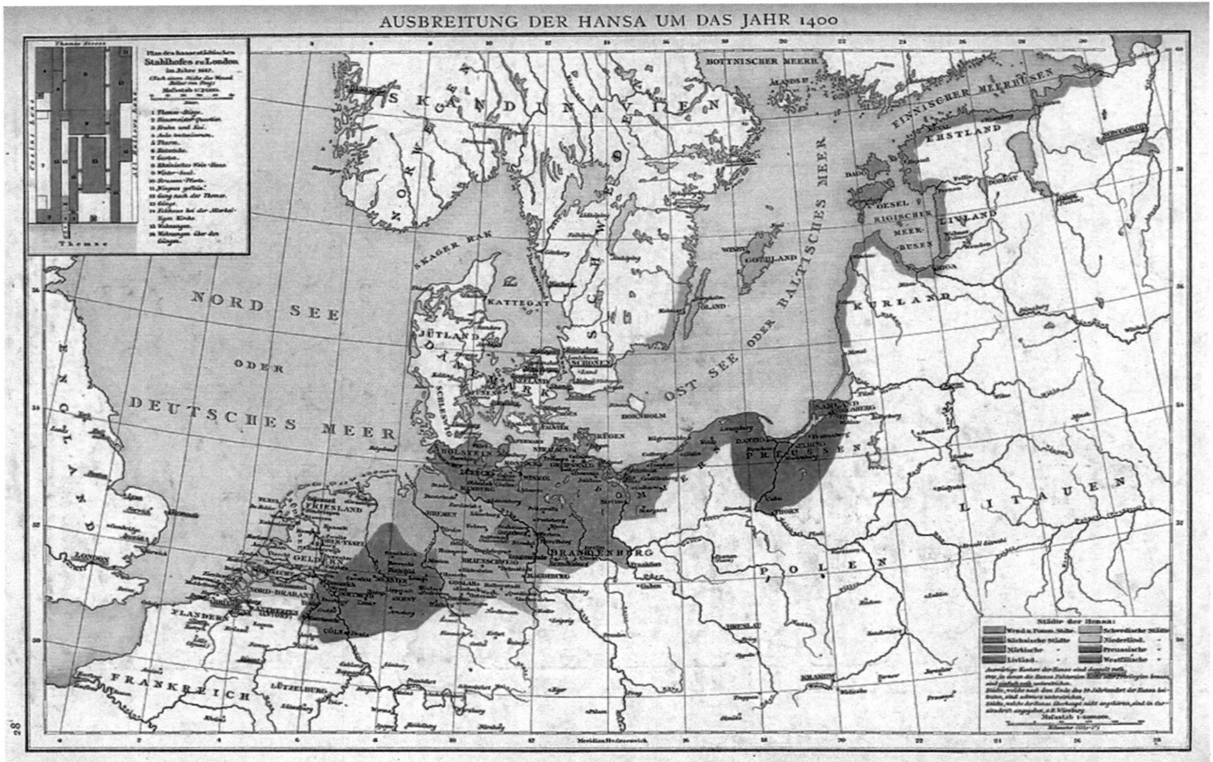

Figure A2.2 Northern Europe in the 1400s, showing the extent of the Hanseatic League.

Source: Plate 28 of Professor G. Droysens, *Allgemeiner Historischer Handatlas*, published by R. Andrée, 1886; *The spread of the Hanseatic League in the year 1400*, Wikipedia. https://commons.wikimedia.org/wiki/File:Ausbreitung_der_Hanse_um_das_Jahr_1400-Droysens_28.jpg (viewed 13/12/2022).

energy source produced by photosynthesis not in our lifetime but hundreds of millions of years ago. The energy-rich, chemically reduced carbon formed by this process had been sequestered in the interim—in underground land-based mines in the case of coal and in a variety of subterranean reservoirs, both land- and sea-based, in the case of oil and natural gas.

The beginning of the coal-dominated era can be traced to Thomas Newcomen's delivery of a coal-fired steam engine to a coal mine in England's Staffordshire in 1712. The steam engine was required to pump water from the mine, allowing recovery of the coal contained therein. Waterlogging was an endemic problem for England's coal mines, reflecting the prevalence of the country's generally wet temperate climate. By clearing the mines of water, the Newcomen steam engine allowed the output of coal to increase exponentially. The improvements to its design, introduced later by James Watt, allowed Britain to emerge as the undisputed home of the Industrial Revolution. Coal would develop as a staple of the British energy economy, finding an ever-widening range of uses: for cooking, home-heating, production of steel, manufacture of textiles, shipping, rail transport, and eventually for the generation of electricity. So great was the power coal bestowed on Britain that in little more than a century after its first major deployment in industry and transport, it would enable this otherwise relatively isolated offshore nation to emerge as the first hegemonic power the world would experience since the demise of Ancient Rome.

The discovery of oil in Pennsylvania in 1859 had a similarly important influence on the economy of the US. The early application of the Pennsylvanian oil was to produce kerosene for use in lighting as a substitute for whale oil. By far the more important subsequent application, however, was the deployment of oil and oil products in transportation—as fuels for cars, trucks, ships, and eventually airplanes.

The widespread use of fossil fuels has allowed much of the modern world to attain a level of affluence and wellbeing that one could not have imagined a century earlier. But the progress has come at a price, and the bill is now due. To maintain our present living standards, we are exhuming at a rate without precedent the carbon buried in the Earth's crust and emitting it to the atmosphere at a rate the planet will have difficulty accommodating, posing a formidable challenge for the continuing, sustainable, life-supporting function of the global climate system.

We turn our attention now in Chapter A3 to a more focused consideration of the conditions responsible for the current, ongoing changes in atmospheric composition. Chapter A4 follows with a discussion of the attendant implications for climate change.

Notes and References

1. Diamond, Jared (1997). *Guns, Germs, and Steel: The Fates of Human Societies.* W.W. Norton & Company.
2. Under the Dead Sea, warnings of dire drought: Lamont-Doherty Earth Observatory, Columbia University, *Science Daily*, 22 March 2017. https://www.sciencedaily.com/releases/2017/03/170322143139.htm (viewed 9/12/2022).

3. McEvedy, Colin and Jones, Richard (1978). *Atlas of World Population History*. Penguin Books.
4. Lott, Dale F. (2002). *American Bison: A Natural History*. Volume 6. University of California Press.
5. Diamond, Jared (1997). *Guns, Germs, and Steel: The Fates of Human Societies*. W.W. Norton & Company.
6. Ibid.
7. In honour of Jean Louis Agassiz, a Swiss-American, latterly Harvard-based, scientist credited over the course of his life with an amazing range of contributions not only to geology and biology but also to glaciology.
8. John Perlin (1989). *A Forest Journey: The Role of Wood in the Development of Civilization*. Harvard University Press.
9. Maria Kielmas: Ancient Sumerian Levees & Canals. *Sciencing*, 2018. https://sciencing.com/ancient-sumerian-levees-canals-16874.html (viewed 13/12/2022).

ANNEXURE 3

Perspectives on the Changing Composition of the Atmosphere

The previous chapter in this Appendix, Chapter A2, described the emergence of humans as a force for global change. As discussed further in Chapter A4, it is apparent that our contemporary presence and influence extends from pole to pole, from the surface to the top of the atmosphere, and even to the depths of the ocean. The best evidence to support this assertion comes from measurements of gases trapped in polar ice, a record of the changing composition of the atmosphere over time.

As indicated in Figure A3.1, the concentration of carbon dioxide (CO_2) has varied between about 190 parts per million (ppm) and 280 ppm over most of the past 650,000 years, higher during the brief warm intervals that punctuate the record, lower during the more extensive intervals of glaciation. Similar fluctuations are observed for methane (CH_4), a variation in this case from about 400 parts per billion (ppb) during glacial conditions to about 700 ppb during the intervening warm periods. The pattern changes, however, abruptly over the most recent few hundred years, as indicated by the data presented using the expanded temporal scale in Figure A3.2. The variations in the concentrations of CO_2 and CH_4, and in fact also nitrous oxide (N_2O), are indicated by essentially vertical lines to the right of the panels included in Figure A3.2. It would strain credulity to argue that natural processes could be responsible for the departure from the trend observed most recently in the record summarized in Figures A3.1 and A3.2: the deviation from the norm is associated with an anomaly present for as little as a few parts in 10 million in time over the entire 650,000-year extent of the record. And besides, there are credible explanations for the unprecedented recent increase in the concentration of these important greenhouse gases: our accelerating use of fossil fuels, specifically for CO_2, complemented for CH_4 and N_2O by contributions from animal husbandry, rice cultivation, and nitrogenous fertilizer.

The contribution of a particular greenhouse gas to climate change may be defined in terms of a quantity referred to as radiative forcing. Radiative forcing provides an estimate for the net change in energy absorbed by the atmosphere associated directly with a specified change in atmospheric composition. The impacts of several of the more important greenhouse gases are summarized in Figure A3.3. The scale to the left refers to a baseline defined with reference to the concentrations of these gases in the atmosphere in 1750 and is intended to underscore the importance for the climate system of the recent rise in the concentration of these gases. The data indicate that as a direct result of the changing concentrations of these gases, the atmosphere is gaining energy at a globally averaged current rate of about 3.2 W/m^2 relative to conditions that applied in the pre-industrial period. This may not seem like much. According to the information discussed in Chapter A1 of the Annexure, it amounts to as little

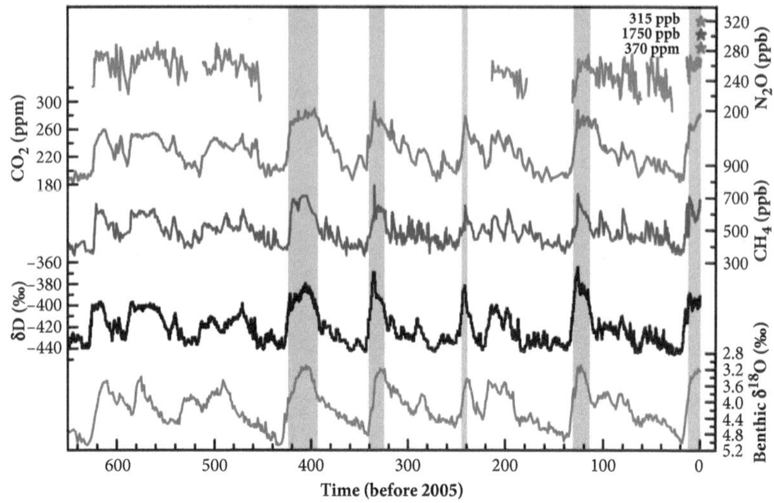

Figure A3.1 Variations of deuterium (δD), a proxy for local temperature, and atmospheric concentrations of greenhouse gases CO_2, CH_4, and (N_2O) derived from air trapped within ice cores from Antarctica and from recent atmospheric measurements.

Source: The Intergovernmental Panel on Climate Change Fourth Assessment Report, 2007. https://www.ipcc.ch/assessment-report/ar4/ (viewed 13/12/2022).

as 0.3% of the net energy absorbed from the sun. The impact is amplified, however, by the consequent increase in the concentration of water vapour. As indicated in Chapter A1, water vapour is by far the most important of the greenhouse gases. The globally averaged surface temperature of the Earth has increased by about 1 °C over the past couple of hundred years. The bulk of this change can be attributed to the increase in water vapour triggered by the ongoing rise in the concentration of the globally, well-mixed greenhouse gases included in Figure A3.3. CO_2 is responsible for about 65% of the net change, with about 16% due to CH_4. The balance is attributed to N_2O, with smaller, declining contributions from chlorofluorocarbons (CFCs), emissions for which were regulated under the Montreal Protocol in 1987. Variations in the concentrations of CO_2, CH_4, N_2O, and CFCs used in the derivation of the results summarized in Figure A3.3 are displayed in Figure A3.4.

We turn attention now to a more focused look at the factors responsible for the current changes in atmospheric composition. In setting the stage for this discussion, it is important to recognize the overriding importance of fossil fuels as the source for the most important of these related greenhouse gases, CO_2. Reflecting the sluggish nature of uptake by the ocean and land biosphere, approximately 50% of the CO_2 emitted by burning fossil fuels since the beginning of the industrial age has been retained by the atmosphere, contributing to the long-term increase in the concentration of atmospheric CO_2. The balance is transferred in approximately equal measures

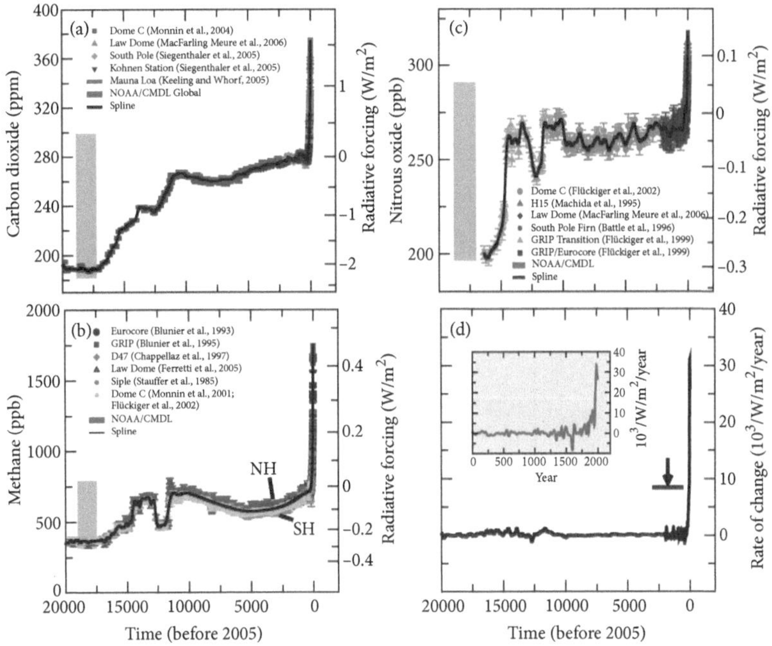

Figure A3.2 Concentrations and radiative forcing by (a) CO_2, (b) CH_4, and (c) N_2O, and (d) rate of change in their combined radiative forcing over the last 20,000 years, reconstructed from Antarctic and Greenland ice and firn data (symbols) and direct atmospheric measurements.

Source: The Intergovernmental Panel on Climate Change Fourth Assessment Report, 2007. https://www.ipcc.ch/assessment-report/ar4/ (viewed 13/12/2022).

to the ocean and to the land biosphere. As indicated earlier, however, the terrestrial sink has been reduced, especially in recent years, by extensive deforestation in the tropics. Consequently, the net contribution of the terrestrial biosphere to the carbon budget is thought at this time to be relatively small. A conceptual representation of the cumulative carbon budget, the inventory of sources and sinks, dating back to 1850 is presented in Figure A3.5. According to this reconstruction, 58% of the carbon emitted from consumption of fossil fuels over this extensive period remains in the atmosphere. The dominant contribution is from coal, followed by oil, with a smaller, though increasing, source from natural gas.

China has emerged as the world's largest emitter of CO_2, surpassing the US for this unenviable position in 2006. The 27 countries of the European Union (EU27) rank number three in terms of emissions, followed by India, the latter rising rapidly to number four. Breakdowns of emissions from the four leading national emitters are presented in Figures A3.6–A3.9. What stands out here is the dominance of emissions from coal for the two large developing countries, China and India, while the mix of

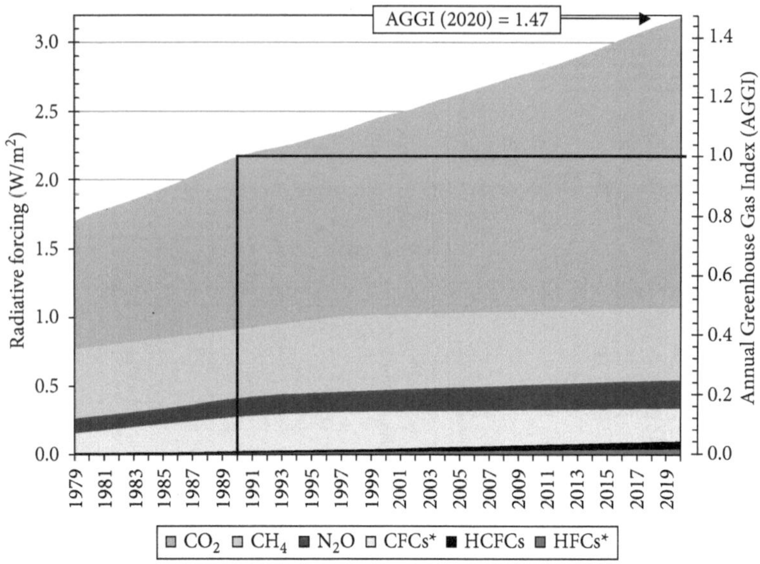

Figure A3.3 Radiative forcing of all the long-lived greenhouse gases relative to 1750. The NOAA Annual Greenhouse Gas Index (AGGI), which is indexed to 1 for the year 1990, is shown on right axis. The AGGI measures the commitment society has already made to living in a changing climate. CFCs, chlorofluorocarbons; HCFCs, hydro-chlorofluorocarbons; HFCs, hydrofluorocarbons.

Source: NOAA Global Monitoring Laboratory, NOAA index tracks how greenhouse gas pollution amplified global warming in 2020. https://research.noaa.gov/article/ArtMID/587/ArticleID/2759/NOAA-index-tracks-how-greenhouse-gas-pollution-amplified-global-warming-in-2020 (viewed 15/12/2022).

sources for the US and EU27 is more diversified. Coal is the fuel of choice for the generation of electricity in both China and India. It enjoyed a similar status in the US up to about 2005, when it was replaced for this purpose primarily by natural gas. It lost out earlier in the EU27: as a result, related emissions of CO_2 have been on a declining trend in the EU since the mid-1980s and again natural gas has been the fuel of choice for the power sector. The shift away from coal in the US and the EU27 was prompted not by concerns about the consequences for climate change, but rather by the need to address problems relating to air quality and associated implications for public health. Oil, deployed primarily in the transportation sector together with a few specific applications in industry, is now the dominant source of emissions of CO_2 for both the US and EU27. Overall emissions of CO_2 have been flat to slightly down over the past several decades in the countries now identified with the EU27 and are beginning to trend down now also in the US. The growth in emissions has slowed recently in China, but is still on a rapid upward trajectory in India.

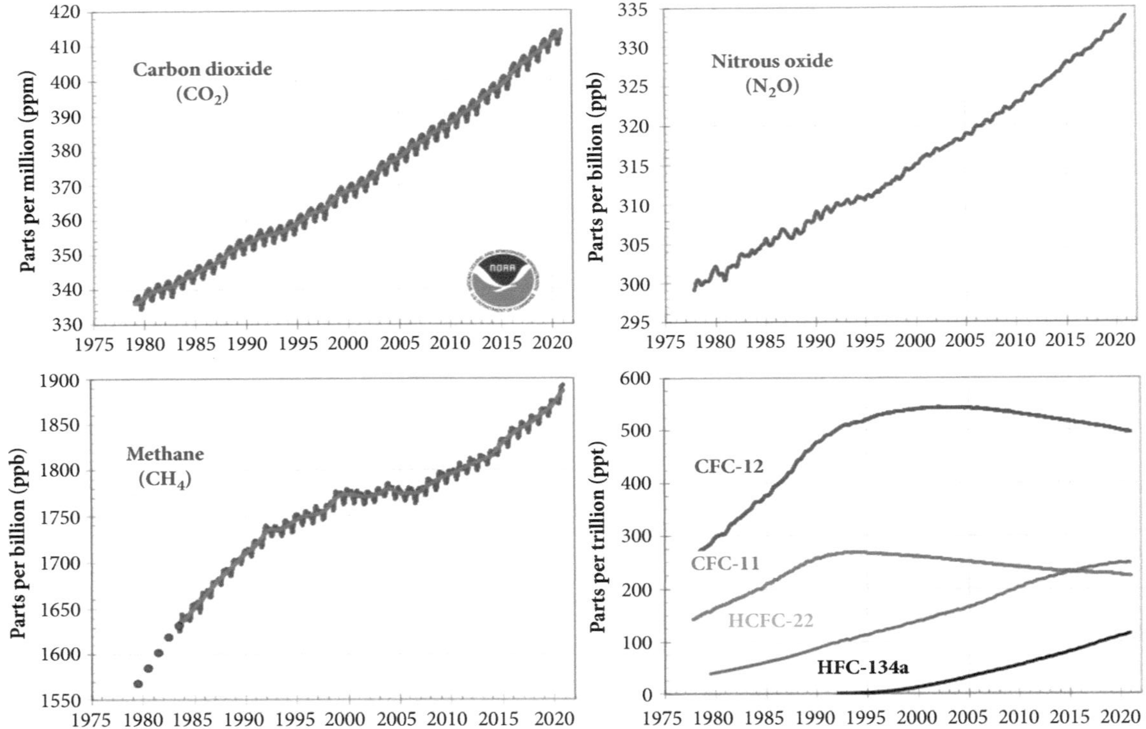

Figure A3.4 Global average abundances of the major, well-mixed, long-lived greenhouse gases— CO_2, CH_4, N_2O, CFC-12, and CFC-11—from the NOAA Cooperative Global Air Sampling Network, with results plotted since the beginning of 1979.

Source: NOAA Global Monitoring Laboratory, NOAA index tracks how greenhouse gas pollution amplified global warming in 2020. https://research.noaa.gov/article/ArtMID/587/ArticleID/2759/NOAA-index-tracks-how-greenhouse-gas-pollution-amplified-global-warming-in-2020 (viewed 15/12/2022).

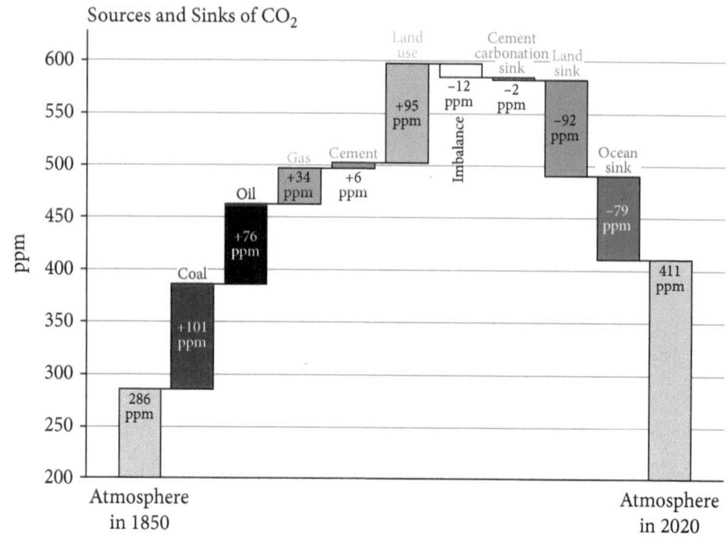

Figure A3.5 Cumulative contributions to the global carbon budget from 1850.

Source: Friedlingstein et al., Global Carbon Budget 2021, 2022.

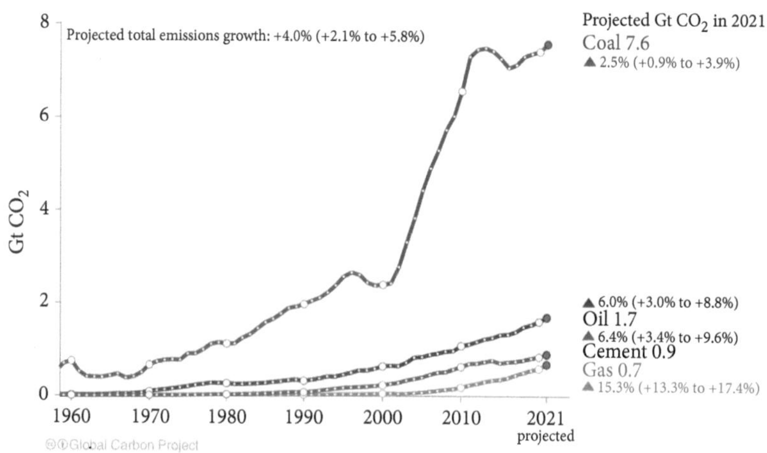

Figure A3.6 Annual fossil CO_2 emissions in China.

Source: Friedlingstein et al., Global Carbon Budget 2021, 2022.

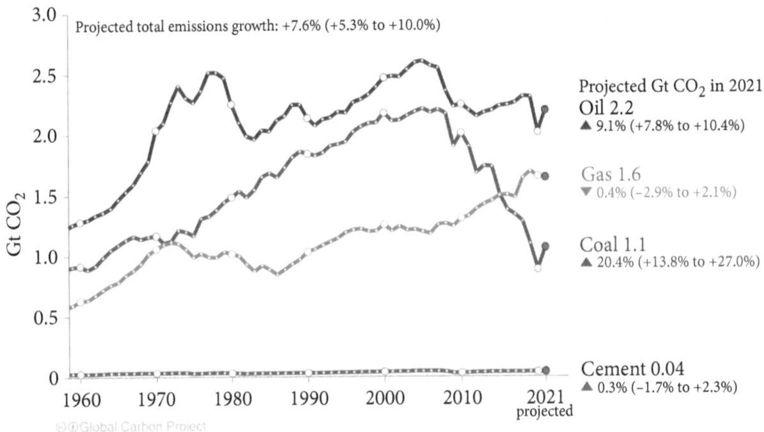

Figure A3.7 Annual fossil CO_2 emissions in the US.
Source: Friedlingstein et al., Global Carbon Budget 2021, 2022.

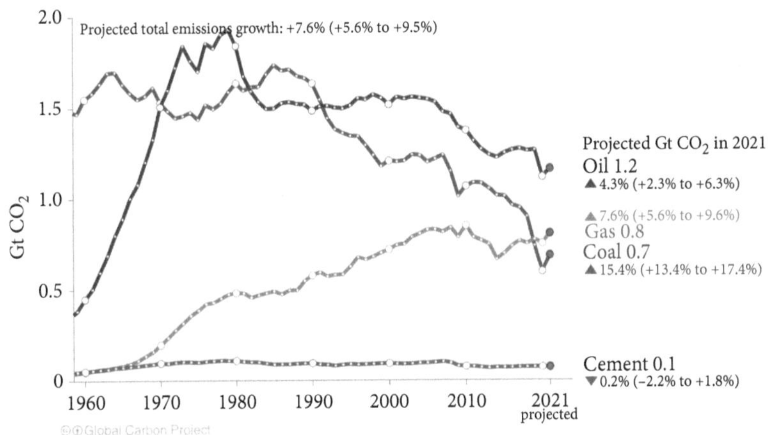

Figure A3.8 Annual fossil CO_2 emissions in the EU.
Source: Friedlingstein et al., Global Carbon Budget 2021, 2022.

On a per capita basis, China is responsible for about half of the CO_2 emissions contributed by the US, 7.4 tonnes per person in 2020 as compared to 14.2 tonnes per person for the US. In the same year, India was responsible for a relatively minor source of 1.8 tonnes per person. Measured in terms of cumulative emissions dating back to 1850, the US is responsible for about 25% of the total, the EU27 for 15%, China for 10%, and India for as little as 3%.

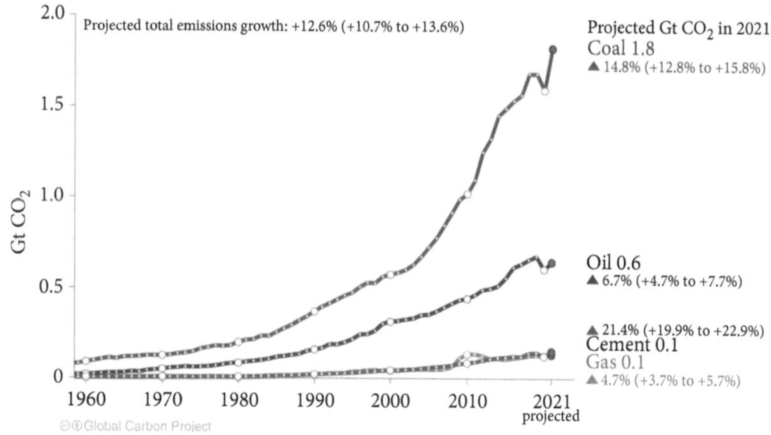

Figure A3.9 Annual fossil CO_2 emissions in India.
Source: Friedlingstein et al., Global Carbon Budget 2021, 2022.

In terms of its impact on climate, CH_4 is the second most important of the globally distributed greenhouse gases, surpassed only by CO_2. On a per unit mass basis, on a 20-year timescale, CH_4 is 86 times more significant than CO_2 in its capacity to alter climate and 28 times more effective over a 100-year period.[1] Its abundance, however, is significantly less than that of CO_2, by a factor of more than 200 in 2019 (1,875 ppb as compared to 410 ppm). The lifetime of CH_4 is relatively brief, however, compared to that of CO_2: less than 10 years compared to hundreds of years for the latter. What that means is that if we were able to cut down on the source of CH_4, or alternatively enhance the efficiency of its removal, we could buy time to adjust to the climate impact of the rising emissions of CO_2.

The concentration of CH_4 has risen significantly over the past several hundred years, from a pre-industrial value of 720 ppb in 1750 to the level of 1,875 ppb observed in 2019. As indicated in Figure A3.4, there have been three separate phases to the trend observed since 1980: a steady increase in concentration from 1980 to about 1990, a flattening of the trend up to about 2007, with a resumption of growth in more recent years. The flattening of the trend from 1990 to 2007 was interpreted initially as an indication that the budget of CH_4 had reached a new steady state, a balance between production and loss. It is clear from the recent record, however, that the source has continued to grow. Concentrations have increased at rates varying from 12.7 ppb per year in 2014 to 10.7 ppb per year in 2019. Concurrent with this increase has been a change in the isotopic composition of CH_4, as indicated in Figure A3.10. Methane has become isotopically lighter. The abundance of the light isotope, ^{12}C, has increased relative to that of the heavier isotope, ^{13}C, over this period. Interpretation of these data promises to offer important clues as to the factors responsible for the recent change in the abundance and temporal variation of the concentration of CH_4 in the atmosphere.

There are two primary sources of atmospheric CH_4. One involves release of CH_4 from mining of coal, combined with leaks associated with the production and

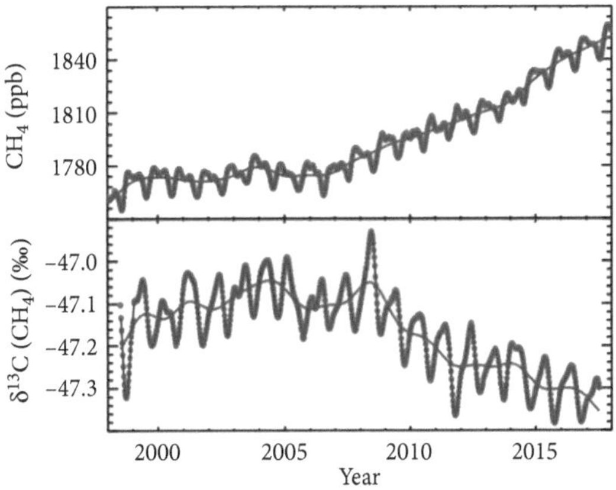

Figure A3.10 Global trend in recent concentrations of CH_4 (upper panel) together with a record of recent changes in isotopic composition of CH_4 (lower panel).

Source: Nisbet et al., Global Biogeochemical Cycles, 33, 318–342, 2019.

distribution of natural gas. The second relates to bacterially mediated decomposition of organic matter in a variety of oxygen-deprived environments, including natural swamps, rice paddy fields, and the stomachs of ruminants (specifically cattle, sheep, and goats), extending to sources associated with disposal of organic waste. The important point is that the two sources differ in terms of their isotopic signatures: the coal/natural gas source is isotopically heavy, enriched in ^{13}C, while the biological sources are typically isotopically light, enriched in ^{12}C. The isotopic record up to the 1980s was consistent with a dominant role for the coal/natural gas source as the primary reason for the upward trend in the concentration of CH_4. The change in the isotopic record since 2007 (Figure A3.10) suggests that biological sources are now more important.

An analysis by Jackson et al.[2] assigned emissions for 2017 to sources as follows: natural wetlands, 39%; agriculture and waste, 38%; fossil sources, 18%; with the balance, 5%, attributed to biomass burning. These results are based on what is referred to as top-down analyses, which seek to use atmospheric measurements as input to an inversion procedure designed to constrain sources, including their geographic distribution. Jackson et al.[3] concluded that three regions—Africa plus the Middle East, China, and South Asia including Oceania—contributed equally to the growth from 2007 to 2017, with a somewhat reduced role for the increase in emissions originating from the US (about half of that attributed to the other three regions). There is a particular need for a focus on studies of sources originating from the tropics, which the authors argue may be responsible for as much as 64% of current global input emissions of CH_4.

The contemporary trend in the concentrations of CH_4 and related emissions is in any event consistent with the conclusion reached by the IPCC in its report addressing

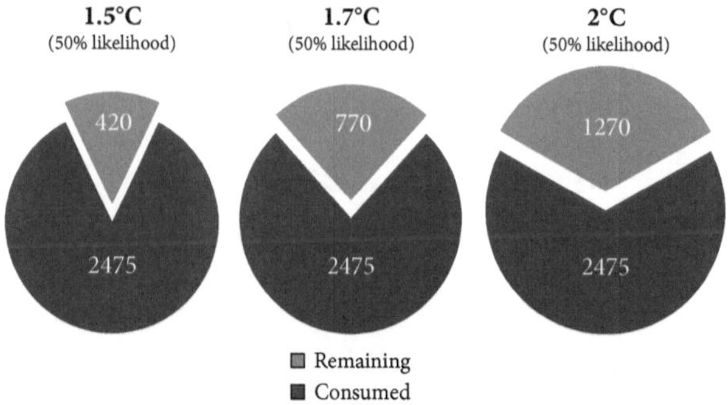

Figure A3.11 The remaining carbon budget to limit global warming to 1.5 °C, 1.7 °C, and 2 °C is 420, 770, and 1,270 Gt CO_2 respectively, equivalent to 11, 20, and 32 years of emissions at the rate that applied in 2022. Since 1750, 2,475 Gt CO_2 have been emitted.
Source: Friedlingstein et al., Global Carbon Budget 2021, 2022.

the target of limiting future growth in globally averaged surface temperature to 1.5 °C or less.[4] Their report concludes that 'limiting warming to 1.5 °C implies reaching net zero CO_2 emissions globally around 2050 and concurrent deep reductions in emissions of non-CO_2 forcers, particularly methane'. To address this objective as it relates to CH_4 will require directed studies to refine our understanding of the relative importance of the various sources. The implications with respect to CO_2 are summarized in Figure A3.11. Meeting the 1.5 °C target will require limiting future emissions of CO_2 to a cumulative total of 420 Gt CO_2, equivalent to 11 years of emissions at the current rate. Figure A3.11 includes also data identifying challenges to meet alternative targets of 1.7 °C and 2 °C: the inferred time windows for these cases correspond to 20 and 32 years respectively. As discussed in subsequent chapters, it will not be easy to meet these objectives. We turn now in Chapter A4 to discussion of the implications of the ongoing climate changes and their consequences for the future.

Notes and References

1. IPCC Fifth Assessment Report (AR5), 2014. https://www.ipcc.ch/assessment-report/ar5/ (viewed 15/12/2022).
2. Jackson, Robert B., et al. (2020). Increasing anthropogenic methane emissions arise equally from agricultural and fossil fuel sources. *Environmental Research Letters* 15(7):071002.
3. Ibid.
4. Rogelj, Joeri, et al. (2018). Mitigation pathways compatible with 1.5 °C in the context of sustainable development. Global warming of 1.5 °C. IPCC, pp. 93–174.

ANNEXURE 4

Perspectives on Earth's Changing Climate

In Chapter A3 we introduced the concept of radiative forcing, a measure of the effect of a specific change in the composition of the atmosphere on the energy budget of the planet. We discussed this in the context of the importance of the so-called well-mixed greenhouse gases. Concentrations of these gases, led by CO_2, have been increasing steadily over the past several hundred years. Their associated radiative forcing is positive, and they contribute consequently to net warming of the planet. There are other human influences, however, that can have the opposite effect, to trigger net cooling, or at least to oppose the warming resulting from the greenhouse gases. Of particular interest in the current context are the changes associated with emissions of sulphur and nitrogen oxides released primarily as by-products of the consumption of coal. Other influences of consequence include production of what is referred to as black carbon—think of soot. In contrast to the negative impact of sulphate, the radiative forcing is positive in that case. Figure A4.1 presents a summary of the multiple contributions from various atmospheric constituents to net global radiative forcing predicted to have developed by 2011, with respect to a baseline identified with 1750.

Chemical reactions are responsible for conversion in the atmosphere of the gaseous forms of sulphur emitted in the combustion process to condensed species identified chemically as aerosol sulphate. This material constitutes an important source of pollution, particularly production of particles of small size (PM2.5) that can penetrate deep into human lungs, posing a serious problem for public health. Sulphate is responsible also for acidification of rainfall, with a variety of negative effects for natural systems receiving this acid input. Our concern here, though, is with the direct impact of the particulate matter on climate and its role as a source of potentially serious negative radiative forcing.

The aerosol formed from sulphate is typically a light-coloured material. As such it can be effective in reflecting sunlight, thereby cutting down on the intensity of solar radiation that penetrates to the planetary surface, a source therefore of negative radiative forcing. This is referred to as the direct influence of the aerosol phase. A more consequential impact of the sulphate material is that it can serve as a medium for the condensation of atmospheric water. If the condensable water is distributed over many condensation centres—a large number of small sulphate aerosols—the net effect is production of a relatively bright cloud. This leads to enhancement of the planetary albedo, with a consequent increase in negative radiative forcing.

This is referred to as the indirect effect of aerosol forcing and is generally interpreted as the more important of the two influences on the atmospheric energy budget.

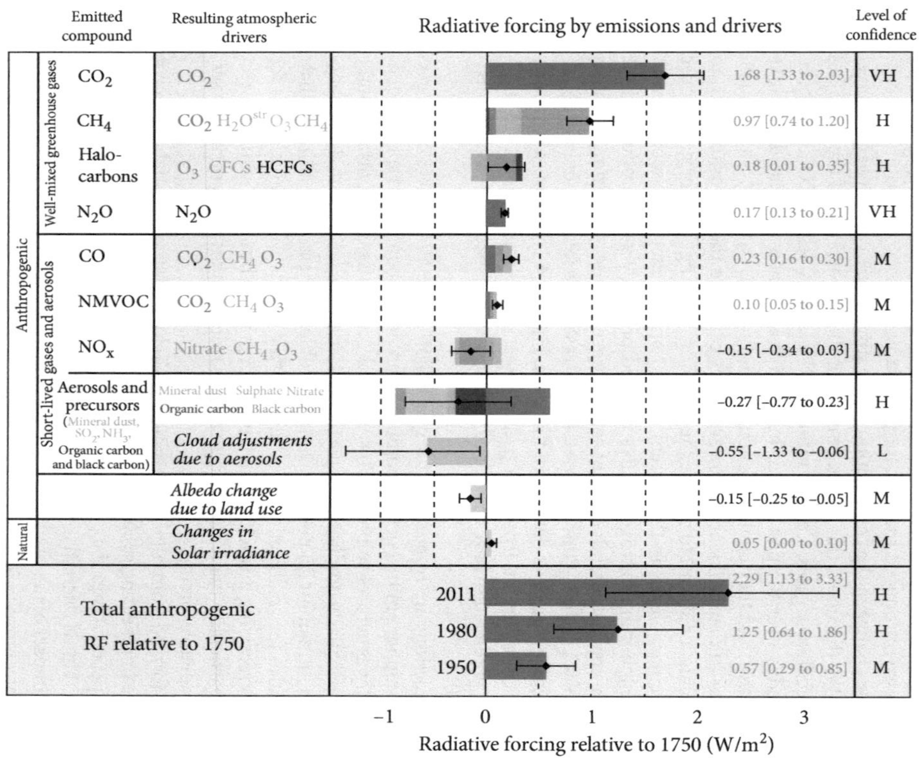

Figure A4.1 Historical radiative forcing for different emissions and drivers relative to 1750.

Source: The Intergovernmental Panel on Climate Change (IPCC) Fifth Assessment Report (AR5), 2014. https://www.ipcc.ch/assessment-report/ar5/ (viewed 9/12/2022).

Understanding the historical climate trend

Just how serious the threat from global warming has become can be judged from the trend in annually averaged global surface temperatures over the 140 years of the existing thermometer-based record presented in Figure A4.2. The US National Oceanic and Atmospheric Administration (NOAA) identified 2020 as the warmest year in the record in terms of land-based measurements, 1.59 °C above the 1880–1920 average. Combining land and ocean measurements, the data presented inFigure A4.2 ranked 2020 as number two, an average of about 1 °C above the long-term baseline. The northern hemisphere was warmer than the globe in terms of land- and ocean-based measurements both individually and in combination. Ten of the globally warmest years are associated with the past 15 years. The five years since 2016 represent the warmest period over the entire modern record. We would need to go back to the last interglacial period, 125,000 years ago, to find a period when average temperature conditions were as warm as those observed over the past decade. This was pointed out in the Technical Summary of the Sixth IPCC Assessment Report (AR6) (2022).[1] Even under the most optimistic conditions, temperatures are expected to rise by at least a further 0.5 °C, more likely by more than 1 °C, over the next few decades. That could bring us back to conditions last realized during the Mid-Pliocene warm period 3.2 million years ago.

Figure A4.3 displays a breakdown in the trends observed over the ocean as compared to the land. Not surprisingly, trends in temperature at the surface of the ocean lag those on land, reflecting the greater heat-absorbing capacity of the former. This is because heat absorbed at the surface of the ocean can be shared over a significant body of underlying water compared with the relatively limited communication of heat to subsurface materials on land. The spatial distribution of the change observed in 2021 is presented in Figure A4.4.

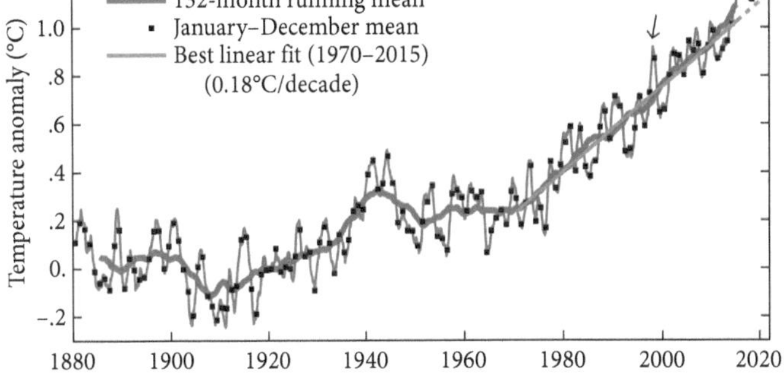

Figure A4.2 Global surface temperature relative to 1880–1920 mean.

Source: Columbia University Earth Institute, Updating the Climate Science, Global Temperature, 2022. http://www.columbia.edu/~mhs119/Temperature/ (viewed 10/12/2022).

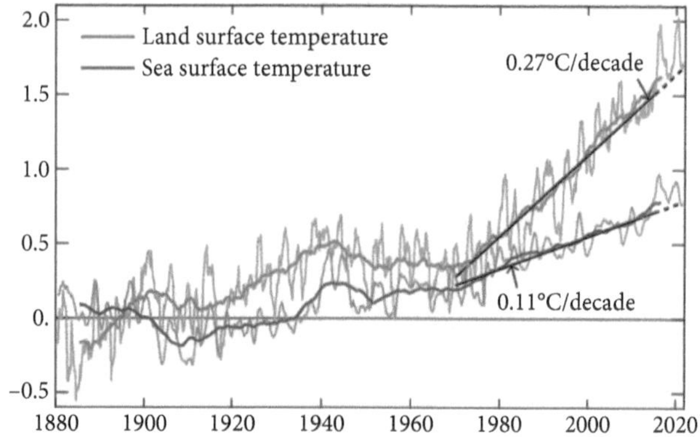

Figure A4.3 Land and sea surface temperatures relative to 1880–1920 mean.

Source: James E. Hansen with Makiko Sato and Reto Ruedy, Global Temperature in 2021, 2022. http://www.columbia.edu/~jeh1/mailings/2022/Temperature2021.13January2022.pdf(viewed 10/12/2022).

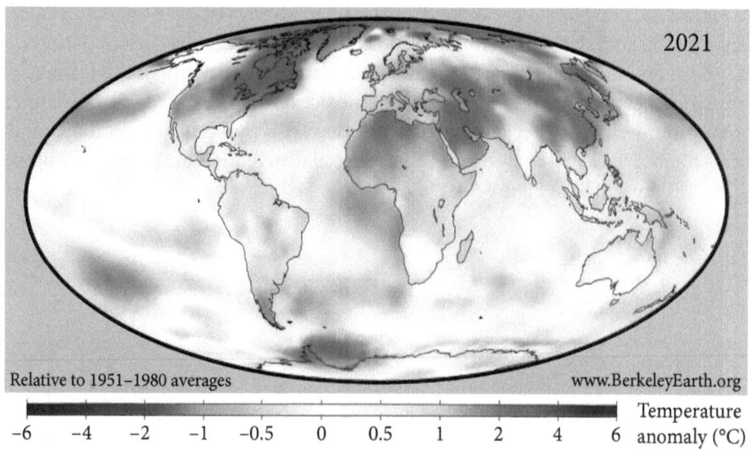

Figure A4.4 Global surface temperature changes in 2021 relative to 1951–1980 mean.

Source: Berkeley Earth, 2021 Global Temperature Changes Since 1850, 2022. http://berkeleyearth.org/dv/2021-global-temperature-changes-since-1850-2/ (viewed 14/12/2022).

The data displayed in Figures A4.2 and A4.3 indicate that there have been three phases in the overall temperature trend: an initial rise between about 1910 and 1940, a hiatus between about 1940 and 1970, followed by a steady increase in more recent years. The warming due to the increase in the concentration of the well-mixed greenhouse gases offers a convincing explanation for the overall rise in global mean surface temperature, an increase of about 1 °C with respect to the 1880–1920 average. The challenge is to account for the interruption of the trend between 1940 and 1970. An increase in emissions of sulphur from accelerated use of coal over this period, combined with conversion to reflective aerosol sulphate which serves to limit penetration of sunlight to the surface offers a plausible explanation for the hiatus.

The years during and immediately following the Second World War were associated with increased economic activity fuelled in large part by coal, with a particular concentration of developments in North America and Europe. A by-product of this activity was an increase in air pollution. An unusual meteorological situation over London in 1952—higher temperatures aloft, a condition referred to as an inversion—caused smoke and foul-smelling fumes to accumulate in the atmosphere over a period of 4 days between 4 December and 8 December. More than 4,000 people died during this episode. A similar incident in Donora, Pennsylvania, in October 1948 resulted in up to half of the residents of that small town becoming sick, with a significant number of associated deaths. The consequence was a widespread call for action both in Europe and the US to address the problem. The solution that governments adopted was not, however, to stop using coal but rather to cut back on the black smoke and the noxious emissions associated with its use. Emissions of sulphur and nitrogen oxides were not addressed until later when public attention switched to focus on problems associated with acid rain.

Reports of extensive fish deaths attributed to acidified precipitation in rivers and lakes in Scandinavia and New England drew important attention to the acid rain issue, as did related incidences of forest dieback in Europe. The result, both in Europe and the US, was extensive legislation to limit emissions of acidic precursors from coal. The transition out of the global surface temperature hiatus can be attributed to the success of this legislation, a conclusion supported by the study by Deser et al.[2] based on analysis of results from a scenario using a global climate model. The importance of the sulphate aerosol effect persists, however, for China and, more notably, as we will discuss later, for India and South Asia.

Impact of anthropogenic climate change

As indicated in Figure A4.4, the impact of human-induced global warming is not distributed uniformly over the planet. Warming is particularly significant at high latitudes in the northern hemisphere, specifically in the Arctic Ocean. The decline in the areal extent of sea ice in this region has been particularly notable in recent years. Sea ice coverage in the Arctic maximizes typically in March, declining to a minimum in September. As indicated in Figure A4.5, the area occupied by sea ice at the September minimum has decreased by close to 50% in recent years compared to the average observed from 1979 to 2000. Given current trends, the Arctic Ocean may be ice free and navigable for a significant portion of the year in the not-too-distant future, offering an opportunity for shipping to proceed unimpeded between Asia and Northern Europe through the Northwest Passage, with important implications for both security (potential US/Russian conflict) and global trade.

A series of mutually reinforcing feedbacks conspire to amplify the impact of warming in the Arctic. A decrease in ice cover results in a decrease in the local albedo,

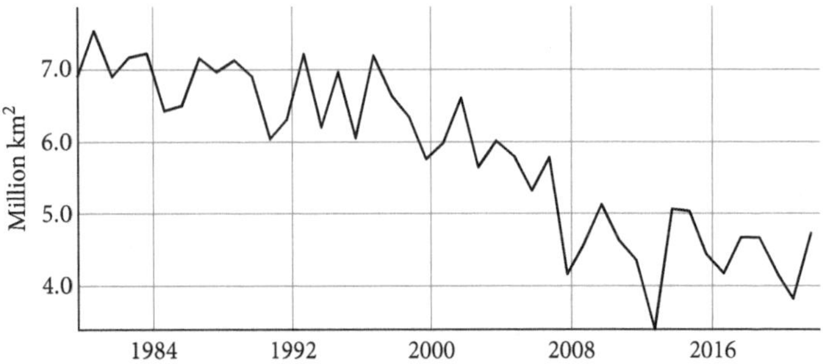

Figure A4.5 Historical Arctic September minimum sea ice extent.
Source: Global Climate Change, Arctic Sea Ice Minimum Extent, 2022. https://climate.nasa.gov/vital-signs/arctic-sea-ice/ (viewed 14/12/2022).

allowing an increase in absorption of sunlight, contributing to a warming of the surface water, triggering a further retreat of the surface ice. The decrease in ice cover has implications for climate overall in the polar region, with a possible extension of impacts to weather at lower latitudes, and potentially also for the circulation of waters in the North Atlantic, which in turn could affect the operation of the climatically important global ocean Conveyor Belt.

Could we expect the amplified warming observed in the northern polar region to show up also in the southern polar region? There is evidence already for a decrease in the ice cover of the ocean bordering Antarctica. It set a record low in terms of its spatial extent in February 2022, coincident with the onset of anomalously warm temperatures over much of the continent, a circumstance that attracted significant attention from polar scientists.[3] The stability of the major ice sheets over Antarctica is clearly a matter of concern, since Antarctica is the depository for the bulk of the world's land-based ice. It is worth noting, while declining to offer this as a forecast for the future, that during the penultimate deglaciation 125,000 years ago when the Earth was as little as a degree or so warmer than it is at present, sea level was as much as 5 m higher than it is today.

Figure A4.6 provides a record of the changes in sea level that have occurred globally over the past 140 years. Sea level over this period has risen by about 24 cm. Contributing factors to this rise include reduction in the volume of water stored in mid-latitude glaciers and expansion of waters in the ocean caused by ocean warming. The contribution due to release of water from the ice sheets in Greenland and Antarctica is currently relatively minor, but there are indications that this contribution may be on an upward trend.

Figure A4.7 provides a record of the fate of the excess heat absorbed by the ocean. The data displayed here were obtained largely as a product of the international Argo float programme,[4] a series of drifting robotic probes deployed essentially globally, designed to record ocean temperatures and salinity to depths as great as 2,000 m. A notable feature of these data is the relatively recent increase in the heat content of the ocean at depth, below 700 m. The ocean is clearly out of equilibrium currently in terms of its heat budget. Understanding how this disequilibrium can impact the

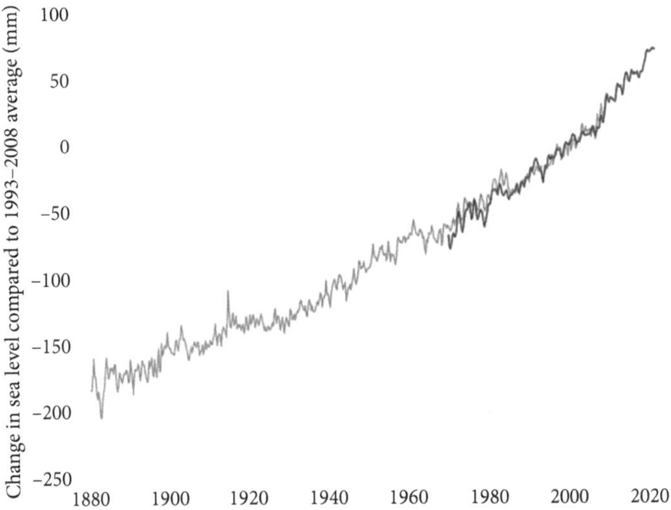

Figure A4.6 Historical global sea level change relative to 1993–2008 average.

Source: Climate.gov, Climate Change: Global Sea Level, 2022. https://www.climate.gov/news-features/understanding-climate/climate-change-global-sea-level (viewed 15/12/2022).

future course of climate, including potentially its effect on Antarctic ice, represents an important challenge for contemporary global climate science. In any event, it is clear, given the accumulation of heat in the deep ocean, that the climate change for which we are responsible will persist for centuries if not millennia in the future even if we were successful in limiting or even eliminating our current emission of greenhouse gases. As the IPCC assessment report AR6 points out, sea level will continue to rise and surface levels of the open ocean will continue to acidify, with a plethora of negative impacts for a host of vulnerable marine organisms.

Weather is becoming more extreme

Discussing the evidence for an increase in the incidence of extreme weather, one of the authors, McElroy[5], led off with a quote from an article by Thomas Friedman that appeared in *The New York Times* on 13 September 2011. Referring to the growing frequency of unusual weather, Friedman posed the question: 'Is it weird enough for you?' He went on to define 'global weirding' as 'the hots get hotter, the wets get wetter, and the dries get drier'. This is a pretty accurate summary of what we have been experiencing at an accelerating pace over the past several decades. The question is why and how it relates to the background issue of global climate change.

The energy of a storm is derived in no small measure from the change of phase of water (conversion of vapour to either liquid or ice). We may assume with confidence that a warmer atmosphere will contain a higher concentration of water vapour.

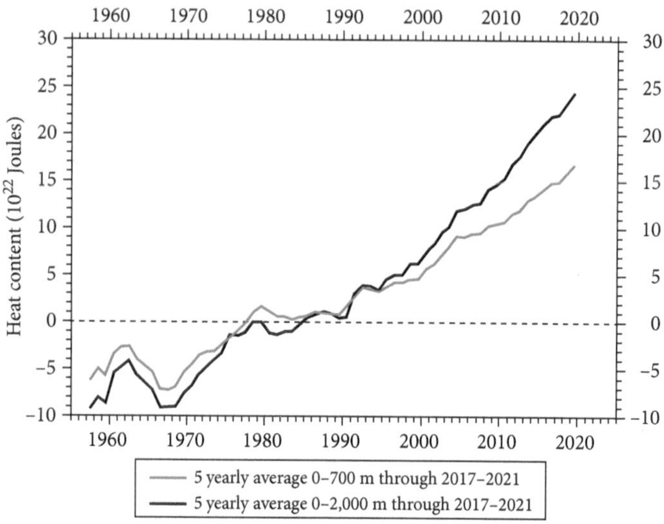

Figure A4.7 Historical global ocean heat content for water depths between 0–700 m (grey) and 0–2,000 m (black).

Source: National Centers for Environmental Information, Global Ocean Heat and Salt Content: Seasonal, Yearly, and Pentadal Fields, 2022. https://www.ncei.noaa.gov/access/global-ocean-heat-content/#null (viewed 15/12/2022).

When conditions favour precipitation—cooling associated with rising motion—the likely result will be an increase in the intensity of the resulting precipitation. On a global basis, precipitation must be balanced by evaporation—demand must equal supply. Evaporation, the ultimate source of the water vapour, is promoted by absorption of energy at the surface, with an important contribution from the ocean. There is no evidence as yet for a major change in overall global precipitation in the face of human-induced global warming. If we assume that the change is minor, what that means is that if we have intensified precipitation in one region, it must be balanced by reduced precipitation elsewhere. The consequence, as summarized provocatively by Friedman, is that the wets get wetter and the dries get drier—increased incidence of floods in some regions offset by droughts elsewhere. Further, given the higher concentration of atmospheric water vapour, we should expect that the storms when they occur could be more intense, fuelled by the enhanced supply of atmospheric water vapour.

Jeff Masters[6] provides an instructive summary of the extremes of weather experienced globally in a single year, in 2021. These included an historic heatwave in western North America in late June, responsible for 808 deaths in western Canada, with an additional 229 in the north-western US. The only deadlier weather disaster in 2021 was extreme flooding in India that claimed 1,292 lives. Hurricane Ida, which made landfall at Port Fourchon, Louisiana on 29 August, was responsible for devastating floods that extended over much of the mid-Atlantic and coastal regions of the Northeast US. The damage in this case, estimated by NOAA, came to $75 billion.

Flooding between 12 July and 18 July in western Germany and eastern Belgium killed 240 people and was rated as the costliest weather disaster in European history. Zhengzhou, China, experienced record rainfall, 64 cm (25.38 inches) over a 24-hour period on 20 July 20, a year's worth in a single day. There were 21 named storms in the 2021 Atlantic hurricane season (third highest on record), 8 of which made landfall in the US, as against the 11 that did so in 2020. And this is only a partial list of the extremes in floods, droughts, and out-of-control fires experienced across the world in the course of a single year.

Masters[7] extended his discussion of the weather disasters observed in 2021 to provide an account of events that developed over the first 9 months of 2022. Quoting the insurance broker Aon from its quarterly disaster report issued on 18 October, he noted that there were at least 29 billion-dollar weather disasters prior to October in 2022. The damage from Hurricane Ian, which devastated Cuba, Florida, and the Carolinas between 27 September and 1 October, is expected to exceed $100 billion dollars. Heatwaves in Western, Southern, and Central Europe killed at least 3,700 people in June and a further 13,000 in July. In addition, these heatwaves caused serious damage to agricultural systems, disrupted cooling systems for power plants, interrupted the supply of power from hydro facilities, and posed problems for shipping on some of the continent's major rivers (the Danube, Rhine, and Po).

Arguably, though, the greatest weather-related disasters in 2022 were those experienced by Pakistan. March was the second warmest month on record in Pakistan, followed by April which broke the long-term record. Temperatures in Jacobabad exceeded 100 °F for 51 consecutive days, peaking at 123.8 °F on 14 May. Rivers swollen by record glacial melt were forced to accommodate the excess of waters resulting from record rainfall during the monsoon season. Pakistan experienced its wettest months since records began in 1961 during July (180% above normal) and August (243% above normal). The floods and torrential rains combined to devastate the lives of the country's citizens, triggering serious shortages of food and lasting damage to the nation's infrastructure.

The circulation of the global atmosphere is changing

There is evidence for a change in what is known as the Hadley circulation, which largely determines meteorological conditions in the tropics and subtropics. Heat absorbed from the sun in the equatorial region causes moisture-laden air to rise as the first step of this circulation system. After losing most of its moisture to precipitation, the air spreads out to higher latitudes, sinking to the surface at latitudes between about 25 and 35 degrees. As the air sinks it warms, by as much as 9 °C per kilometre of descent. Having lost most of its moisture in the rising portion of the circulation, the air when it returns to the surface is not only very hot but also very dry. The major desert systems of the world are associated generally with the descending branches of this circulation. There are indications that the Hadley circulation has expanded in response to this warming world, by as much as a degree of latitude between 1979 and 2005.[89] Think of the consequences of such a change: the Sahara Desert could extend across the Mediterranean into southern Europe and the south-western desert of the US could move north, threatening the integrity of the grain-producing region of the country. And all this can be triggered by what might appear as just a minor adjustment in the circulation of air in the tropics.

We noted earlier the enhanced warming underway in the polar region, including notably the Arctic Ocean. A consequence of this warming is that the gradient of temperature between mid and high latitudes in winter is now on a downward track. The gradient in temperature sets up a gradient in pressure that results in a river of air that circles the planet at a latitude of about 60 degrees, the Polar Jet Stream. The reduction in the temperature gradient leads to a decrease in the speed of the air associated with this circulation system. The consequence is that the air in the Jet Stream can begin to wobble, looping to the south in some regions with compensating loops to the north elsewhere. If you are on the upwind side of a southern loop, you can be bathed in semitropical air. On the other hand, if you are on the downwind side of a northern excursion, you could experience air that just a day or so earlier could have achieved its sub-zero temperature from passage close to the North Pole. The net effect is episodes identified with conditions of unusual warmth offset by periods of unusual cold. The US experienced one of these extremes on the cold side between 12 February and 20 February in 2021. This was associated with heavy snow, freezing rain, and severe cold in Texas and surrounding states. As many as 246 people lost their lives and the associated damage was estimated at an all-time winter storm record of $23 billion. A week of unusual weather in the US Midwest in mid-December 2021 offered an example of an excursion to the warm side. That system spawned 69 tornadoes across western and central Kentucky, followed just 5 days later by a massive cyclone that whipped across regions as far north as Wisconsin and Minnesota, conditions unprecedented for that time of year.

Climate change and air quality implications for India

The climate of India is controlled in large measure by seasonal changes in the difference of temperatures between the land and surrounding ocean, what is referred to as a monsoonal circulation. The principal features of the winter and summer properties of the monsoon are illustrated in Figure A4.8. India receives the bulk of its annual precipitation during the months of July to September from the south-west phase of the monsoon which is driven by the relative warmth of the land compared to the surrounding ocean. The pattern reverses during the winter months, January through March. During this period, dry temperate air from the north-east is warmed adiabatically as it descends over the Himalayan mountain chain. The winds pick up moisture from the Bay of Bengal, depositing this as rain in Southern India and as precipitation when the moist air rises over the hilly mountain chain known as the Eastern Ghats.

Analysis of an extensive temperature record dating back to 1901 indicates that India has warmed significantly, especially since about 1980, with the rise evident not only in annual records but also in data for individual seasons.[10] What stands out, however, in this extensive record is an anomaly in the behaviour observed in the north-east extending into the south-central contiguous region. The authors make a convincing case that the anomaly is associated with the influence of a thick blanket of particulate matter developed as a result of industrial activity, particularly in the north, with contributions potentially also from south-western China, compounded by emissions associated with burning of agricultural and municipal wastes. This brown haze,

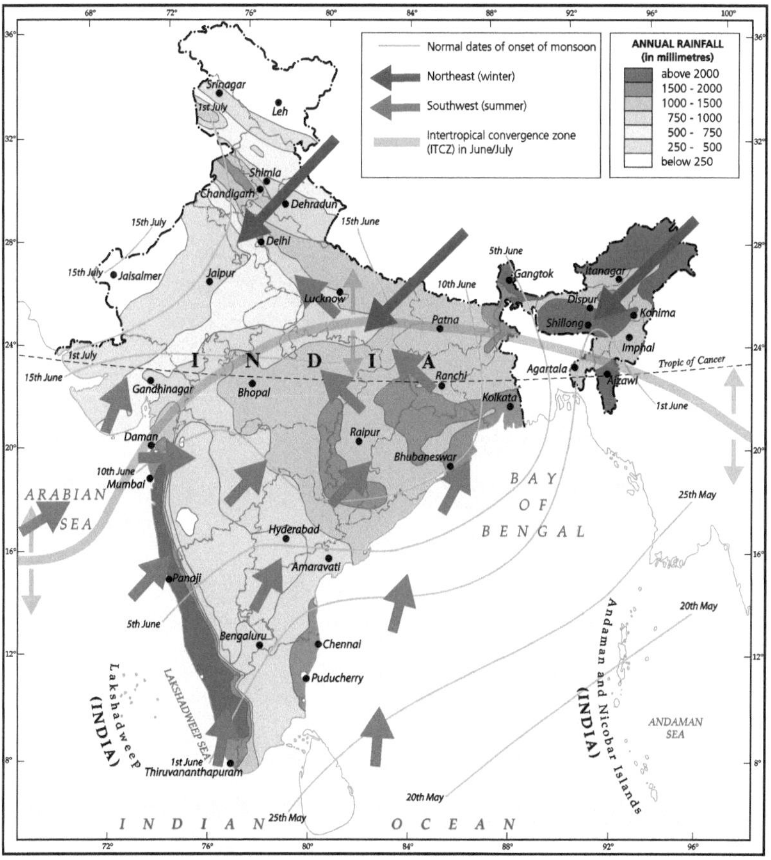

Figure A4.8 Onset dates and prevailing wind currents of the south-west and north-east monsoons.
Source: Oxford University Press, 2024.

they suggest, absorbs a fraction of the incoming solar radiation, cutting down on the energy that would otherwise reach the surface, accounting thus for a lag in warming in the impacted region.

The situation bears some resemblance to the conditions discussed earlier for the US and Europe in the post 1940 period. Action must be taken in the interests of public health to reduce this pall of pollution in India. The significance of the problem is illustrated by data from the IEA[11], summarized in Figure A4.9, which indicate that, more often than not, air quality, especially in northern regions of India, is unacceptable in terms of its impact on public health. We discussed the issue of air pollution in more detail in Chapter 6. Addressing the problem is likely to alter the properties of the Indian monsoon, with implications potentially not only for the intensity but also for the spatial distribution of precipitation. A report from the India Meteorological

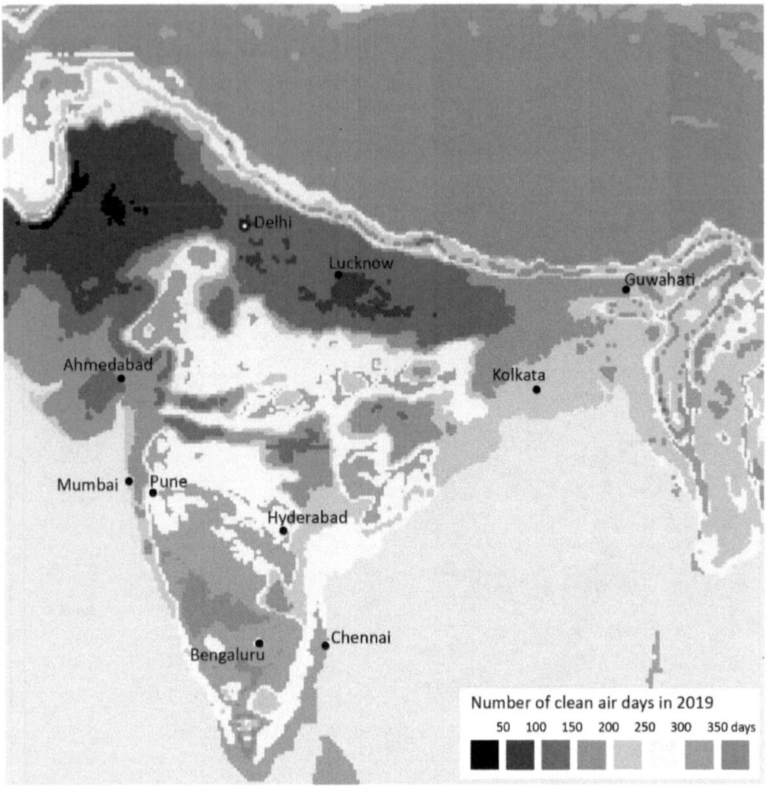

Figure A4.9 Number of clean air days, 2019.
Source: Blue Sky Analytics (2020) and IEA (2021), India Energy Outlook 2021. https://www.iea.org/reports/india-energy-outlook-2021 (viewed 16/12/2022).

Department[12] indicates that the impact can be subtle. In terms of nationwide rainfall, the Department classified the 2021 south-west monsoon as normal. They went on to qualify this statement, however, as follows: 'considering the month-to-month rainfall variation over India as a whole, the season is very uniquely placed in the historical record for its distinct and contrasting month-to-month variation'.

Notes and References

1. IPCC Sixth Assessment Report (AR6), 2022. https://www.ipcc.ch/assessment-report/ar6/ (viewed 9/12/2022).
2. Deser, Clara, et al. (2020). Insights from Earth system model initial-condition large ensembles and future prospects. *Nature Climate Change* 10(4):277–86.

3. Antarctica sea ice concentration hit record low, *The Washington Post*, 2022. https://www.washingtonpost.com/weather/2022/03/18/antarctica-sea-ice-record-low/ (viewed 14/12/ 2022).
4. Argo. https://argo.ucsd.edu/ (viewed 15/12/2022).
5. McElroy, Michael B. (2016). *Energy and Climate: Vision for the Future.* Oxford University Press.
6. Jeff Masters: The top 10 global weather and climate change events of 2021, Yale Climate Connections, 2022. https://yaleclimateconnections.org/2022/01/the-top-10-global-weather-and-climate-change-events-of-2021/ (viewed 15/12/ 2022).
7. Jeff Masters: World rocked by 29 billion-dollar weather disasters in 2022, Yale Climate Connections, 2022. https://yaleclimateconnections.org/2022/10/world-rocked-by-29-billion-dollar-weather-disasters-in-2022/ (viewed 16/12/2022).
8. Fu, Qiang, et al. (2006). Enhanced mid-latitude tropospheric warming in satellite measurements. *Science* 312(5777):1179.
9. Seidel, Dian J. and Randel, William J. (2007). Recent widening of the tropical belt: evidence from tropopause observations. *Journal of Geophysical Research: Atmospheres* 112(D20).
10. Ross, Robert S., et al. (2018). Decadal surface temperature trends in India based on a new high-resolution data set. *Scientific reports* 8(7452).
11. Blue Sky Analytics (2020); and IEA (2021), India Energy Outlook 2021. https://www.iea.org/reports/india-energy-outlook-2021 (viewed 16/12/2022).
12. India Meteorological Department: Salient Features of Southwest Monsoon 2021. https://mausam.imd.gov.in/Forecast/marquee_data/Presentation_Salient%20Features%20of%20Southwest%20Monsoon%202021.pdf (viewed 16/12/2022).